张 鹏 举

适应·更新·生长

内蒙古工业大学建筑馆改扩建设计

ADJUSTMENT | RENOVATION | ENLARGEMENT

The Reconsruction and Extension of the Architectural
Hall of the Inner Mongolia University of Technology

Zhang Pengju

中国建筑工业出版社
CHINA ARCHITECTURE & BUILDING PRESS

序
Foreword

序

工业遗产的改造与利用，已成为近年来人们关注的热点，但我看了一些实际项目后，思想上疑问颇多。工业遗产的保护和利用究竟应该如何把握？似乎有不少问题需要思考和总结。

最近，在朋友的推荐下，有机会参观、体验了张鹏举先生设计的内蒙古工业大学建筑馆（以下称建筑馆）——一个工业遗产保护和利用项目，不少问题在这里得到了解答，很受启发。

我一直在想的第一个问题是，是否凡是XX年代的工业遗产都需要保护？结合目前实际情况，我以为：确定改造项目需要慎重研究，研究原有建筑的条件与新的功能需求是否能真正统一起来。从这个角度看，建筑馆项目的改造，让我看到了一个很好的实证，张鹏举先生和他的团队在确定这个项目前做了细致的、实事求是的调查研究，从原有建筑在校园中的位置、建筑规模、空间特点、结构及设备条件，以至立面的典型性等各个方面，论证了把一个20世纪60年代建造的工业厂房改造成一个有特殊使用要求的建筑馆的可行性，为建筑馆改造的成功，打下了一个极好的基础。比起我看到的那些缺乏理性思考、为炒作而"保护"，甚至以打着"保护"之名行新建、扩建之实的项目，设计人的态度和最后产生的效果，是不可同日而语的。

第二个问题是如何保护。对于这个问题，我一直也有很多疑问，对照那些架房叠屋，"整容"过度的项目，我认为工业遗产改造的重点应该放在对原有建筑的充分开发和利用上。恰恰是这一点，建筑馆做得很成功。我们看到，不仅原有厂房的空间和主体结构保持原状未作改动，而且很多构件如钢柱、铁梯、红砖墙面、水泥地面以至废弃的机器设备，如烟囱、冲天炉、通风管道、设备基础等人予以保留，并结合新的功能和空间要求加以利用。特别是室内不用空调，而是利用原有地道组织室内通风，在一定程度上达到了冬暖夏凉的效果。这种不是简单的改造，而是在改造中对原有建筑充分尊重的态度，十分值得称道。

当然，工业遗产的保护不是被动的，在利用的基础上，如何"更新、创造、再生"，肯定是一个建筑师的愿望，这里有个"适体"和"适度"的问题。建筑馆之所以成功，就是这两点把握得很好。在功能更新的基础上，新老空间、材料、细部和色彩，那么自然地结合在一起，让你感受到一种异样的氛围；开敞而又围合、混杂而又统一、雅致而又质朴、粗犷而又精确，即使改造后形成的一些小空间，如茶室、室外平台等也都能给你一种自然松弛的感受，而这种感受是在一般新建筑中无法体验的。抓住了工业遗产保护的特点从而做出自己的诠释和创造，我想这正是建筑馆之所以成功，并能打动那么多参观者的"奥秘"所在吧。

感谢张鹏举先生给了我一次对工业建筑遗产保护利用问题的体验和解惑的机会，祝张鹏举先生和他的团队创作出更多的优秀作品来。

程泰宁
中国工程院院士
2011-07-12

Foreword

The transforming and utilizing of industrial legacy have become the focus of people's attention in recent years. However, when I have seen some projects that had been put into practice, a lot of uncertainties have immerged. How can we protect and make use of the industrial legacy? It seems that there are many issues to think about and conclude.

Recently, with the recommendation of a friend, I have the opportunity to visit personally the Architectural Hall of Inner Mongolia University of Technology (referred to as Architectural Hall below) designed by Mr. Zhang Pengju— a program of industrial legacy protection and utilizing. Plenty of problems have been worked out here and it is very inspiring.

The first question that has been puzzled me is if the entire industrial legacy left in the year of XX needs to be protected. Associate with the current situation, I think: it needs to consider very carefully what kind of program can be transformed and also needs to have the further study whether the original building condition and the new function can be integrated. From this point of view, the transforming program of the Architectural Hall roles us a very good model. Mr. Zhang Pengju and his team have made a lot of meticulous and practical research and study. The feasibility of transforming an industrial workhouse in 1960s into an architecture house of particular functions has been proved from many aspects as its position on the campus, the coverage scale, the spacious characteristics, the structure and equipment condition, the typical features of its facade, and so on. It lays a very good foundation for the success of the Architectural Hall. Compared with the newly-built and extended-built projects under the name of protection without any rational consideration, the designer's attitude and the final consequence are not easily distinguished from each other.

The second question is about how to protect. I have many doubts toward this issue. Compared with the frame house duplication program and those that seem to have experienced the plastic surgery, I think the industrial transforming focusing point should be placed upon the exploiting and utilizing of the original buildings. The Architectural Hall has been doing a very great job at this point. As we can see, on one hand, the space of the original workshop and the body structure have remained untouched. On the other hand, many parts of the building from the components like steel columns, iron ladders, the red brick wall, the cement floor to some abandoned machinery and equipment as chimney, cupola, ventilation ducts, basic equipments etc, are all kept as they were and utilized according to the new function and space requirement. The issue of keeping warm in winter and cool in summer has been solved through the original tunnel functioning as the ventilation ducts, so that the cost upon the air conditioners is saved. This process is not a simple transforming process, because the designer has been holding an attitude of respect toward the original building throughout the process—this is very commendable.

Of course, the protection of the industrial legacy is not passive, and how to realize the process of updating, creation and regeneration on the basis of making use of the old building is the wish of every architect. There is an issue of appropriateness and moderation. The key to the success of the Architectural Hall is to deal with these two points very well. On the basis of updating the function, the old building and the new building's space, materials, details and colors are combined together very harmoniously, so that you may feel a very exotic and peculiar aesthetical atmosphere here, open and enclosed, mixing and unifying, elegant and simple, rough and exact. Even the small space formed after transformation as tearoom, the outside platform etc. can make you feel relaxed and natural. And you can seldom experience such feelings in the ordinary new building. Seizing the features of the industrial protection and making our own explanation and creation, I assume these are the secrets for the Architectural Hall to be successful and why it can touch so many visitors.

Thanks Mr. Zhang Pengju for giving me such a good opportunity to experience personally the fantastic example of protecting and utilizing the industrial architectural legacy and the questions that puzzled me for a long time have been solved finally due to this visit. I wish Mr. Zhang Pengju and his team can create more and more excellent works.

Cheng Taining
Academician of China Engineering Academy
2011-07-12

前言
Preface

前言

一栋建筑物一经建成，排除战争、自然力破坏或人为拆除等因素，其自身的老化是一个相当缓慢的过程，它可以跨越不同的社会形态而存在。在建筑物漫长的生命周期内，人的生活观念和方式都会随着社会的发展而改变，而建筑物需要适应不同时期使用者的需要，因此，它必须随着社会、政治、经济、文化等因素的发展以及人们生活质量的提高而对其功能和所属的环境进行适应、更新，进而生长。从这个意义上说，建筑物生命周期的全过程应是一个不断适应新需要、不断更新其自身空间环境，并在此基础上有机生长的过程。

目前，由于世界全球化趋势带来的推动力和20世纪末城市产业结构转型所带来的变动，我国很多传统的工业区陷入了困境，大量的产业建筑面临着被拆除的命运。在这种由工业时代走向后工业时代的新陈代谢中，如何对待现存的旧产业建筑是一个现实而又值得思考的问题。值得庆幸的是，国内外已出现了许多优秀的改造案例，在局部区域甚至显现出某种热潮，如上海的8号桥、北京的798等，在此背景下，我们在2008~2009年间经历了一次有意义的旧厂房改造活动。

在内蒙古工业大学校园的中部有一座建于20世纪60年代末的废旧厂房。这组厂房是依据当时生产线建造的铸造车间，在当时，它的生产活动曾一度构成了地区行业的支柱产业。产业结构调整后，车间各部门陆续闲置，到1995年，整体处于废弃状态。2008年初，学校决定改造这组车间，因而成就了这次伴着思考与创作的改造过程。

该组车间的用途最终确定为建筑馆，而其限额限时的改造要求决定了设计过程与建造过程合二为一的特点，同时，建筑师需"兼职"工长、预算员、材料采购员。正是这个非常的经历让设计过程有了更多的感悟。

本书立足于人文与生态视野，以内蒙古工业大学旧厂房改造实践为着手点，解析其设计理念与改造策略，文章尤其着眼于记录改造过程的点点滴滴，希望对同类实践具有参考意义。

全书核心内容分三个部分：一是"适应"，介绍在原有旧厂房空间内如何识别空间、寻求关系、引导出适合改造后使用的新功能——建筑馆；二是"更新"，展示建筑馆新功能布局诱发新场所并完善其机能的过程；三是"生长"，即建筑馆改造完成后二期扩建设计的内容，介绍了整体建筑从纯粹空间走向复合空间的过程。

需要指出的是，本书是2011年《适应·更新·生长——一次人文与生态视野下的旧产业建筑改造实践》的再版。在核心内容不变的基础上增加了建筑馆使用过程中的状态和扩建完成后的内容。

重编再版需要感谢扎拉根白尔的编辑、田忠山老师和王欣欣的英文翻译、刘洁老师的学生画整理和文字以及莫日根、孟一军、张广源、曹扬、陈溯、方振宁、李鹏等的摄影；继续感谢范桂芳老师和同事张恒、薛剑、苍雁飞等在设计过程中的工作和努力。本书仍然使用了赵辰教授、王兴田教授、黄居正教授的评品文章，引录了高旭老师、白丽燕老师的相关文章，同时摘引了一次项目品谈中同行关于建筑馆的评价，在此一并感谢。

Preface

Once architecture was being built, besides the factors of the wars, natural destruction and human dismantlement, aging itself is a rather slow process which could exist across various social patterns. Throughout its long life, people's living concepts and lifestyles change with the social development, whereas the architectures need to adjust to the users in different periods. Therefore, its functions and surroundings must be adjusted, upgraded and developed, going with the development of the society, politics, economy, culture and other factors. In this sense, an architecture's whole life should be in the process of adjusting to the new needs continuously, upgrading its space and surroundings endlessly, and based on that, developing itself exuberantly.

Nowadays, with the motive force by the worldwide globalization and the change of urban construction transformation in late 20th century. In China, many traditional districts are stuck and plenty of industrial buildings confront with the fate of being dismantled. Undergoing the process of metabolism from the Industrial Age to the Post-industrial Age, it's a practical and thought-provoking question how to handle the old existing industrial building groups. Luckily, there are many excellent cases in and abroad and shows like a tide in some places, such as the 8th Bridge in Shanghai, 798 in Beijing, etc. On this basis, we carried out a significant transformation of old factory building from 2008 to 2009.

There is a deserted factory, built in the late 1960s, in the middle of the campus of IMUT. This group of buildings was the workshops according to the previous production lines, at that time, it had been the pillar industry in Inner Mongolia for a long time. After the reorganization of industrial structure, every department of the workshop was stopped one after another and the whole was deserted by 1995. At the beginning of 2008, the university decided to reorganize these workshops, which accomplished the process of thought and creation.

The final and ultimate usage of these workshops is confined as an architectural hall. It features the combination of the process of designing and constructing due to limited time and budget, meanwhile, the architect should also play the roles as the leader, budgeter and material purchasing agent. Such a unique experience made the designing process more unforgettable.

This book roots in the views of humanity and ecology, focuses on the practice of an old factory workshop in IMUT, analysis its designing notion and transformation strategies, especially the specific processes of the transformation. We look forward to bringing some reference to the similar practice.

This book is composed mainly by three parts. First, adjustment. It introduces how to recognize the space and find the relations in old factory workshops so as to elicit the new functions after transformation, that is, to build into an "architectural hall"; Second, improvement. It shows that the new functions and the development to a improve. Third, development. That is the reorganization introducing the process transformation from a simple space to a compound space.

It's worth pointing out that this book is the second edition of the book *Adjustment · Improvement · Enlargement — An Old Factory Renovation in Views of Humanity and Ecology* published in 2011. On the basis of the unchangeable and fixed core content, the book is added with the situation in the using process and the extension content of the Architectural Hall.

The second edition is contributed to many friends: Zhalagenbaier as the editor, Tian Zhongshan and Wang Xinxin as the translators, Liu Jie as the transcript of the students' pictures and words, Morigen, Meng Yijun, Zhang Guangyuan, Cao Yang, Chen Su, Fang Zhenning, Li Peng ,etc. as the photographers. I really appreciate again for the whole designing work of Fan Guifang, Zhang Heng, Xue Jian, Cang Yanfei, and so on. The book still uses the comment articles of professor Zhao Chen, professor Wang Xingtian and professor Huang Juzheng, quotes the related articles of teacher Gao Xu,Bai Liyan, meanwhile, presents critiques of peers' comments on a Project Talk , to whom I also owe many thanks.

目录

序	003
前言	007
一、关于建筑馆的改扩建	013
适应　空间引导功能	018
更新　新功能诱发新场所	024
生长　从纯粹走向复合	096
二、关于建筑馆的思考	115
从校园里的旧厂房开始	116
一次务实的旧厂房改造与再利用实践	122
三、关于建筑馆的评品	129
赵　辰　建筑学的力量	130
王兴田　"逆设计"	139
黄居正　空间：记忆的装置	143
高　旭　从表皮到表情	146
白丽燕　空间的回溯与期待	150
四、建筑馆的画和话	157
建筑馆的画	158
建筑馆的话	170
五、建筑馆的前世今生	177

Contents

Foreword	003
Preface	007
I. The Reconstruction and Extension of the Architectural Hall	013
ADJUSTMENT Space Directs Functions	018
RENOVATION New Functions Stimulate New Places	024
ENLARGEMENT From Simple to Compound	096
II. Thought of the Architectural Hall	115
From the Old Workshop on Campus	119
A Practice of Practical Rebuilding and Recycling of an Old Factory Building	125
III. The Evaluation of the Architectural Hall	129
Zhao Chen Power of Architecture	134
Wang Xingtian Reversing Design	141
Huang Juzheng Space: Mechanism of Memory	144
Gao Xu From Skin to Expression	148
Bai Liyan Retrospect and Expectation of Space	152
IV. The Pictures and Utterances of the Architectural Hall	157
The Pictures of the Architectural Hall	158
The Utterances of the Architectural Hall	170
V. The Past and Present Life of the Architectural Hall	177

一、关于建筑馆的改扩建
I. The Reconstruction and Extension of the Architectural Hall

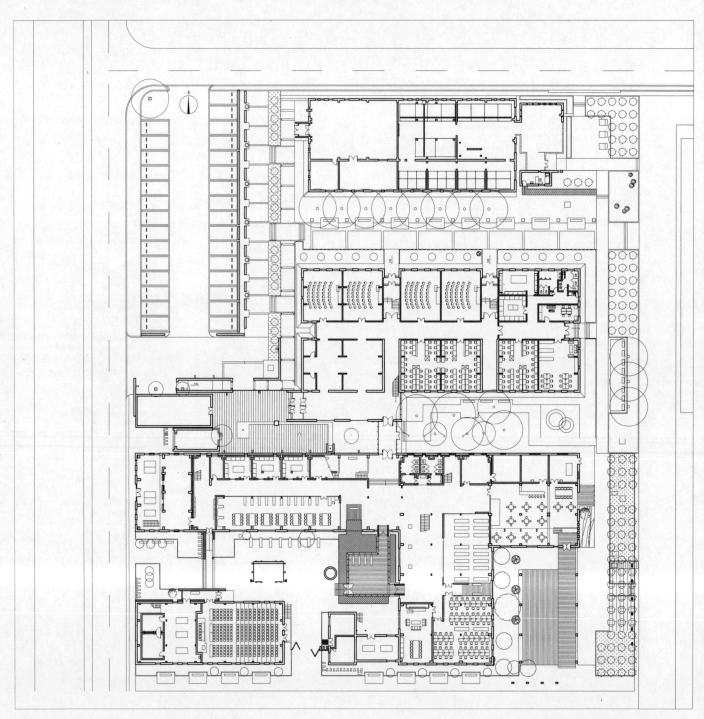

一层平面图（苍雁飞、李登钰 绘）
First Floor(Drawn by Cang Yanfei、Li Dengyu)

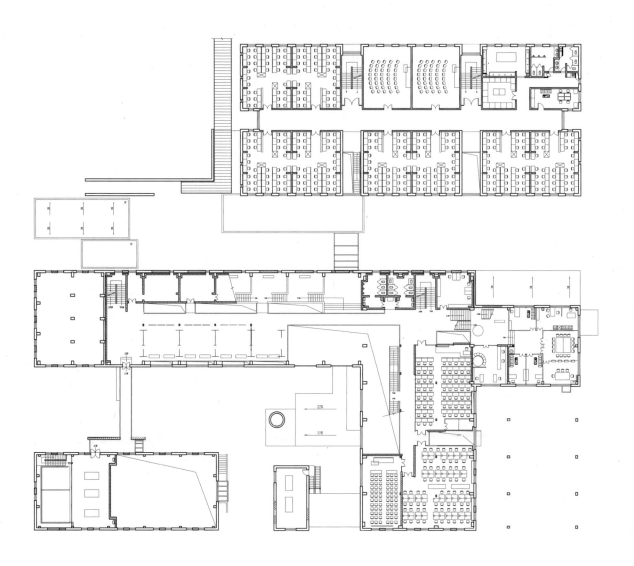

二层平面图（苍雁飞、李登钰 绘）
Second Floor(Drawn by Cang Yanfei、Li Dengyu)

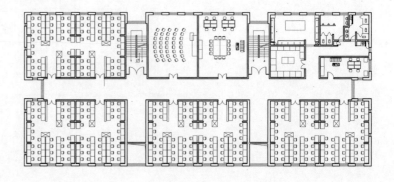
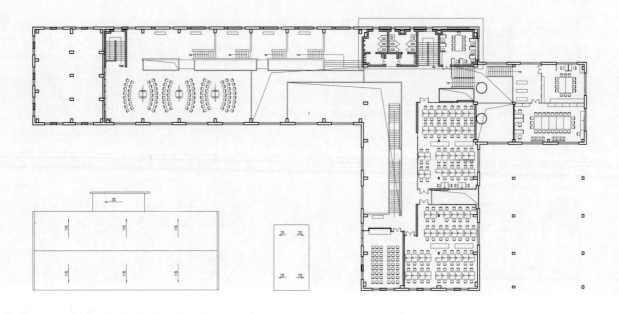

三层平面图（苍雁飞、李登钰 绘）
Third Floor(Drawn by Cang Yanfei、Li Dengyu)

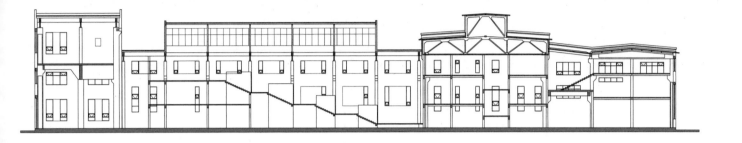

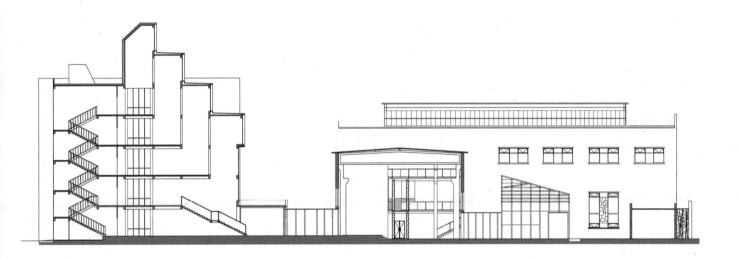

剖面图（苍雁飞、李登钰 绘）
Section(Drawn by Cang Yanfei、Li Dengyu)

适应 空间引导功能
ADJUSTMENT   Space Directs Functions

不论是限制性改造的被动思考还是"绿色"思想的主动行为，建筑师首先要做的是与学校一起考虑这组旧车间新生命的适宜定位。其次，要在已有的每个空间内外体验，为其量身定制一个恰当的具体功能。换句话说，我们的创作活动不只是为某项功能创造一个空间，更多的时候是识别现有空间特征，引导新功能，这正是基于生态视野下的一种自觉行为。这样的行为不同于以往的设计活动。通常，建筑师的工作是为某一项功能创造适宜的空间，但在既有空间的前提下，创作思维常常需要转化。当下，旧产业建筑改造再利用受到广泛重视，因而，这种思维的转化有助于我们在生态与人文的视野下，制定适宜的设计策略。

Whether restraint passive thought or ecological active action, it is vital for the architect to consider with the university the appropriate orientation which can make this old workshop alive. Then, to experience every present space in and out so as to fulfill the proper and detailed function, in other words, our work is not to create a space for the certain function, most of the time, identifying the features of present space and finding out the corresponding usage and function. This process differs in the conscientiousness of ecological view from the previous ones. Previously, the architect created a space for the certain function. But since we have the space first, the thought should be changed. Presently, great focus is put on the transformation and reuse of the old building groups. Thus, such change of thought will do good to design the better and more suitable strategies in view of humanity and ecology.

1 从校园文化活动中心到建筑馆

设计伊始，呈现在脑海的是北京的798、上海的8号桥、纽约的SOHO区、伦敦的Tate博物馆等国内外优秀的改造案例，它们创造的化腐朽为神奇的新功能成为设计初期我们直接效仿的榜样。

改组旧车间最初的定位是校园文化活动中心。当然，这样一种考虑也是基于学校缺少这类人文空间，但很快发现，文化活动中心的功能不能很好地契合厂房空间的通透与开放，其音、体、美等不同的功能构成需要互不影响的空间，以保证不同使用者不会受到彼此的干扰，而过度的分隔又违背了充分利用原空间的"绿色"初衷。因此，我们开始为厂房寻求更为适宜的归宿。此时，正值学校建筑学专业迎接全国教育评估，专业教学条件亟待改善。我们立刻意识到，铸工车间通透开敞的大空间、自然裸露的结构构件、不加掩饰的结构细部都能够与建筑学重交流、重体验、重实践的教学特点相适应，我们认定，这将是一个天然的建筑馆。接下来的工作重点是分析并体验现有的各个空间，将建筑馆的教学、展览、评图、实验、图书等功能一一"对号入座"。

1 From Culture Centre to Architectural Hall

At the beginning, the excellent cases of transformation in and abroad leave deep impression to us, such as, 798 in Beijing, 8th Bridge in Shanghai, SOHO District in New York city, Tate Museum in London, etc. Their amazing and outstanding function set good examples for us to learn from then on.

The initial usage of reorganizing workshops was confined as the campus culture center. Certainly, such place was in shortage; however, its function wasn't fit for the broadness and openness of the workshops. The culture center had many usages: music play, P.E. contests and art shows. All of them need isolate space to avoid disturbance of each other. Besides, dividing the space to a larger extent also went against the initial idea of originality and ecology. Hence we began to pursue a better fate for it. It was in coincidence with the evaluation of architectural engineering in China and the condition of teaching architecture need improving. We realized that the spacious place, exposed components and uncovered details of the workshop can be best suitable to the characteristics of teaching architectural engineering that is, focusing on interaction, experience and practice. We believe that it would be a natural architectural building for sure. The following working points are to have analysis and experiences of every space, fulfill the function of teaching, exhibiting, mapping, experimenting and housing books correspondingly.

2 从空间的开放到功能的开放

无疑，十几米高的开敞车间，引导我们以一种开放的方式布置建筑馆的功能。在"对号入座"之前，一项工作是必先完成的，即加层，从而获得更多的使用面积。加层方案在利于结构加固和功能摆放的前提下进行，并且，设计还始终遵循着一条原则，即在更有利于视野通透的位置加层，积极营造水平无阻、上下通透的开放视野。当然，这种视野的通透也必然造就了光和空气的流通，而它们三位一体的流通则又进一步加强了交流空间的质量，同时也实现了"绿色"的初衷。在竖直方向，为消除高度带来的隔阂感，把展览空间与楼梯相结合，各单元展览平台顺楼梯渐次抬升，有机联系了各层的高度，淡化了楼层感。

在上述策略的基础上，依照空间特点安排功能进行"对号入座"：顶部靠近天窗的部分自然成为一个天然的美术教室；安静的角部设置图书阅览室；南部独立的车间尺度得当，无疑是一个视线和音质俱佳的报告厅；西侧，作为原铸工车间生产线起始端的高大厂房内悬挂着若干巨型倒四棱锥的沙漏，地面上摆放着雕塑感极强的大型设备，俨然是一个完美的模型室……当空间必须分隔时，除需要极为封闭的个别房间使用砖墙外，其余均采用玻璃围隔，并以其所需封闭的程度，依次为透明玻璃、单层U玻、双层U玻、贴膜U玻。

2 From Open Space to Accessible Functions

It's no doubt that the workshop was very large and more than 10 meters in height, which guided us to assign the functions in an open way. Before fulfilled and assigned functions correspondingly, one job must be done first, that is, put into more layers so as to gain more used place. The plan was carried out on the basis of consolidating the structures and fulfilling the functions, on the other hand, from the beginning to the end, followed one rule of being convenient to look through, trying positively to create the far-reaching horizon of no limitation horizontally and vertically, of course, the well-illumination could certainly cause the purpose of ecological idea. In vertical, in order to diminish the isolation feeling due to height, we combined the exhibit space with the stairs. Each exhibit unit was increased in line with the stairs, which connected the height wholly and lighten the feelings of layers.

According to the above scheme, we assigned functions correspondingly space by space, on the top close to the skylight formed into a natural arts classroom. The quiet corners were used as reading rooms. In the south, the workshop's independence and proper size made it into an auditorium which had best vision and sound effect. In the west, because it was originally the place to begin the working line, it was very high and wide, hanging some giant inverted pyramid hourglasses and displaying some large equipment. Because of strong feeling about the strong feeling about the status, it looked as if a perfect mould room. When the space need dividing, besides a few used bricks, all the other used glass of transparent, single-layer glass U, U double-layer glass and U glass paste film to the degree of sealing.

改造中的建筑馆内部

（张鹏举 摄）

The Inner of the Architectural Hall in Renovation

(Taken by Zhang Pengju)

南立面部局（曹扬 摄）
The Layout of the South Vertical Section (Taken by Cao Yang)

3 从"酷"的空间到"酷"的功能

厂房的结构构架和设备机器都传达出了特别的空间气氛。无论是认同其十足的"酷"意，还是出于唤醒历史记忆，设计中总是不忍将其破坏，甚至为它们更加合理的存在，而在任务书中增设了新功能。

东侧的小车间，两个"冲天炉"立于中部，室内管道纵横，梁柱斑驳，颇似某类"藏酷"空间，于是设计将其改造为以水吧和陶吧为主的艺术沙龙；其外的侧院原是带有天车的露天输料场，两排高大的立柱颇有工业味道，具有很好的场所感，经过改造成为校园文化的一个园地和休闲场所；车间的煅烧锅炉，将其表面的维护砖墙和内部的耐火砖剥离，金属构件暴露于外，在"锅炉"内砌上几层台阶，在院落中成就了一处独特的交流场所……

此外，车间中的一些特殊空间还引发我们去尝试创造具有生态意义的新功能。如，院中高耸的烟囱，用管道连于室内，形成了无能耗的通风系统；工程启动后，获知一条贯穿室内各处的地下生产线通道，设计利用其进一步完善了通风系统，做法是：打开室内界面，并联通室外，夏日里，经由地下的凉爽新风在打开屋顶天窗后所形成的气压的引导下流向开放的室内各处，形成天然的空调系统……

3 From Cool Space to Practical Functions

Both the structures and the machines of the workshop send the spectacular atmosphere. Whatever its recognized coolness or recalling the memory, we can't bear the destruction in designing, even we add some new functions to its reasonable existence.

In the small eastern workshop, to cupolas locate in the central point, there are many pipes and posts in it just like a loft room which made it change into an art saloon major in water and leisure bars. The outside yard was originally a conveying plant with overhead travelling crane, and two rows of erective columns have strong sense of industry and space which made it change into a garden and refreshment houses. There are some burning boilers, too. We separate the surface protective brick wall and inner refractory bricks. Thus, the metal constructions were exposed, with many stairs in it made it change into a unique communication site.

Furthermore, some special places of the workshop stimulate us to try out some more ecological functions. For example, the tall chimney was connected with pipes indoors. Forming an incompetent consumption draft system. After the project was carried out, we knew that there was an underground production line throughout every place. Then we used it to improve the draft system. Out steps are as followed: open the indoor interfaces and connect outside. In summer, led by the air pressure caused in opening the skylight of the roof, the new fresh wind from the underground flow to the open rooms everywhere which forms a natural circulation system.

改造前的厂房（张鹏举 摄）

The Factory Before Renovation (Taken by Zhang Pengju)

"煅烧锅炉"内的交流场所（莫日根 摄）
The Communication Site in the "Calcinations' Boiler" (Taken by Morigen)

更新 新功能诱发新场所
RENOVATION　New Functions Stimulate New Places

事实上，一个旧建筑空间如果没有后续适宜的行为活动，它的改造将成为空壳，而促使建筑空间再利用的后续行为不可能提前发生，这需要对新功能进入既有空间的活动进行准确的判断和理性的分析，即对旧空间引导出的新功能再度诱发的新场所进行思考。可见，新功能是促成旧空间改造再利用的催化剂和潜在动力，而产生的新场所质量则需要审慎处理。

设计中对于各种功能接口进行重新塑造，旧的空间被穿越，新的空间被重新界定起来，不同功能或接口在或清晰或含糊的逻辑中显现着相互的联系，进行着或平行或交叉的陈述与对话：

旧厂房的红色砖墙重组后本身就是天然的画布；原有窗户依据其功能所需，用厂房中的锈钢板加以遮掩；厂房中主要的结构都有意裸露，将这些原来建筑中平凡的历史沉淀重归于公共活动的活跃角色；钢、木、玻等新元素的加入，反衬了建筑原有的肌理感和神韵，新旧的交织更加彰显活力；水的引入、玻璃房的镶嵌、新墙体的竖立，加之建筑馆的原有表皮和纤柔的草地重写了这一历史建筑的生气；光线的组织、苍郁老树的保留和开放空间的进一步梳理，把不同功能的接口以创造故事场景的方式融合在一起，使改造后的建筑馆在某种序列中展现着特定的气质和场所氛围。

If an old building space was not followed with afterward appropriate behaviors, the transformation was meaningless. Because those behaviors can not happen ahead of time, the accurate judgment and reasonable analysis of the new functions should be put into the new existing space, that is, consider carefully of the new space caused by the new functions from the old space. Therefore, the new functions are the catalyst and the potential power to reuse the old space and we should bear in mind that coping with the quality of the new place cautiously.

When we design, we recreate the functions of every interface. The old space was passed through and the new space was redefined. Various functions or interfaces are connected in the clear or vague logic and are conducting a parallel or cross statements and dialogues.

After reorganized, the old red brick walls are natural canvas. The former windows are hidden by the stainless steels according to the new requirement. All the major structures of the workshop are uncovered purposely aiming at exposing the common history of building precipitation to the active roles in public. The elements of steel, wood and glass contrast with the original sense of texture of charm, also the interweave between the new and the old highlight the liveliness. Water-introduction, glass house-embedment and new wall's erection combined with the existing epidermis and soft meadows rewrite the life spirit of the historical building. Lights-choosing, trees-preserving, openness-carding and other scenes of the story acculturated with different interfaces of the functions convey the specific temperament and surrounding atmosphere of some sequence in the Architectural Hall.

夏日阳光下的南侧入口局部（莫日根 摄）
Part of South Entrance Side in the Summer Sunshine (Taken by Morigen)

1 新场所的动线节点

当人们在建筑空间中活动时即产生了动线,对于教学空间,多数人的活动会有明确的目标——或教或学或管理,因此,在传统的教学空间中,短捷的动线往往是考虑这类空间质量的一个指针,但在建筑馆这样立足于交流的开放空间中,仅此是不够的。因而,动线中的节点成为关注的目标,它们是交流的场所,是故事的发生地,同时也是另一类"教"与"学"的场地。在此,线中有点、点中有线,甚至线即是点、点即是线。院子是点、入口是点、中厅是点、楼梯是点,当线中有了"桥","桥"也变成了点。

1 The Motive Joints of the New Place

When people act in an architectural space, the fixed lines formed. As for the teaching space, most people can have clear aims: teaching, studying or managing. Therefore, in traditional space, short and efficient lines are one indicator to evaluate such space's quality. However, it is far more enough for an architectural building which aims at interaction. Thus, the motive joints become the focuses; they are the places for communication, for story, meanwhile, another kind of teaching and studying. Here, the lines and spots coexist with each other, even can substitute each other. The spots are yards, entrance, hall and stairs, when the bridges are in the line, they become the spot also.

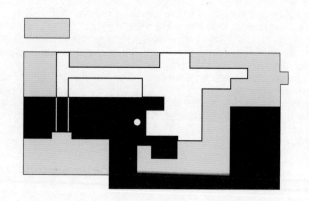
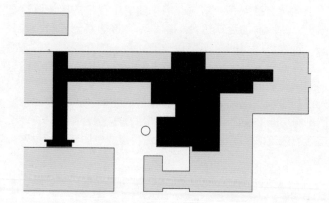

建筑室内动线空间(苍雁飞 绘)
The Circulation Space Inside of the Building(Drawn by Cang Yanfei)

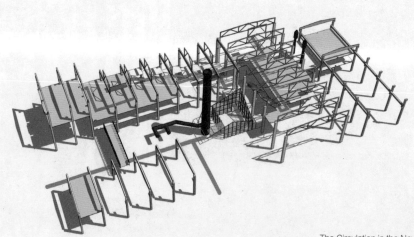

新场所的动线(苍雁飞 绘)
The Circulation in the New Sites(Drawn by Cang Yanfei)

1.1 院子

院子是中国传统建筑中重要的组成部分之一。依照传统，改造后建筑馆的院子自然构成了其空间的精华所在，同时，它也是整体空间动线序列的开端。建筑馆的院子包括东、中、西三个，东部的院子连接着可以通往艺术沙龙的入口，中部和西部的院子则是去往建筑馆主入口的必经之地。

1.1 Yard

Yards are one of the most important parts in Chinese traditional architecture. According to the tradition, the re-built yards naturally composed the essence of the building, at the same time, it's the starting point of the whole space fixed lines. The yards of the building are the eastern, central and western ones. Eastern one joins the entrance to the salon, whereas the other two yards are the necessary way to the main entrance of the building.

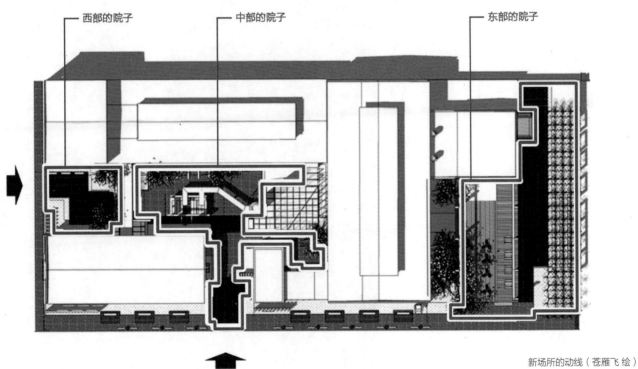

新场所的动线（苍雁飞 绘）
The Circulation in the New Sites(Drawn by Cang Yanfei)

改造后建筑东侧的院子(曹扬 摄)
The Eastern Yard after Renovation(Taken by Cao Yang)

东部的院子

东部的院子在旧厂房中原为配置吊车和输料的场地，是一个边院，改造前，框架裸露、杂草丛生。此外，经年累月的修路抬高了周围的路面，形成了一定的下沉。此院树木浓郁，尺度感良好。

改造的工作是处理边界、完成围合、营造中心场所感和领域性空间。

具体做法是：周边设界墙，辅以摆放机器装置；用台阶过渡院内外的高差，台阶可坐并成"看台"；台上植树阵形成荫凉；框架梁所围处设木板平台，构成中心场地，其北接沙龙入口，天气好时可成为休息喝茶的场所，同时，结合围合的"看台"形成"舞台"，成为一个校园文化的阵地，当然，更是一处方便的室外教学场所。

Eastern Yard

Traditionally, the eastern yard is the place for putting cranes and delivery. It's a side yard and full of exposed frameworks and weed infestation. Besides, many roads have been repaired over years, which enhanced the height of outside road and caused some submerging. There are many luxurious trees in it and it has a very good sense of scale.

The work for transformation includes handling the sides, completing surrounding, creating the sense of place and field.

The detailed actions are: building the wall around with some machines against it; using the steps to transit the high difference, also the steps can be the stands; planting trees outside to provide shade; creating the platform, the central place, at the place of framed girder. It joins the entrance to the salon in the north and can be the place for rest and leisure on nice days, meanwhile, the surrounding stands join together into a stage and create a campus culture center, of course, it's a convenient outdoor classroom.

莫日根 摄
Taken by Morigen

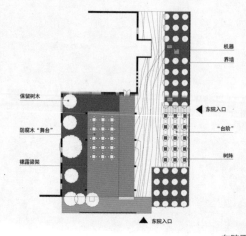

东院平面解析（苍雁飞 绘）
The Structure Diagram of Eastern Yard(Drawn by Cang Yanfei)

方振宁 摄
Taken by Fang Zhenning

夜晚从"舞台"看沙龙入口（曹扬 摄）
From the "Platform" to the Entrance to the Salon in Evening(Taken by Cao Yang)

东部院子局部（莫日根 摄）
Part of Eastern Yard(Taken by Morigen)

东部院子局部(莫日根 摄)
Part of Eastern Yard
(Taken by Morigen)

中部的院子

中部的院子原是旧厂房的核心区，多股人流汇经此处。依据新功能的动线，需要将其改造成进入建筑馆的前院，成为师生们出入场馆的必经之地。

院中保留的烟囱、锅炉及几株大树成为组织空间的关键要素，同时它们也是界定更小空间的关键要素。设计仅是用铺地加以引导，引入水池来切分，一座小木桥跨过水池向室内延伸，在模糊界限的同时，强化"进入"的感受。中心处的锻造锅炉，拆出围护砖后，成为院中的"亭子"。

Central Yard

The central yard is the core of the old factory in the past and groups of people assemble here. According to the fixed line of the new function, we need to transform it into the front yard of the Architectural Hall and the necessary part in and out.

The preserved chimney, boiler and trees are the major components of the new space, also they classify the sections of the smaller space. Our designing is to guide by bricks on the floor and to divide by the pool. There is a small wooden-bridge crossing the pool and reaching indoors, which vagues the boundaries yet emphasizes the sense of entering. The burning boiler, having been pulling down the outside bricks, turns to be the pavilion of the yard.

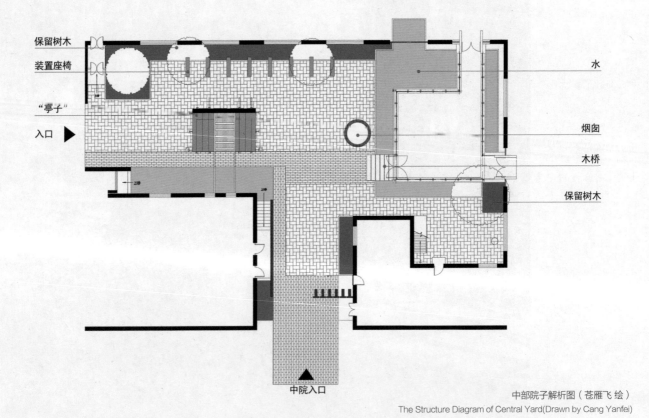

中部院子解析图（苍雁飞 绘）
The Structure Diagram of Central Yard(Drawn by Cang Yanfei)

中部院子局部（莫日根 摄）
Part of Central Yard(Taken by Morigen)

中部院子局部（莫日根 摄）
Part of Central Yard(Taken by Morigen)

中部院子局部(曹扬 摄)
Part of Central Yard(Taken by Cao Yang)

中部院子秋韵（莫日根 摄）
Autumn Melody of Central Yard(Taken by Morigen)

门厅、烟囱、水、光影（莫日根 摄）
The Entrance Hall, Chimney, Water, Shadows(Taken by Morigen)

傍晚从中部院子看门厅（曹扬 摄）
The Hall Seen Through Central Yard in the Evening(Taken by Cao Yang)

南侧院门局部（莫日根 摄）
Part of South Entrance(Taken by Morigen)

进入报告厅台阶（莫日根 摄）
Steps to the Lecture Hall
(Taken by Morigen)

西部的院子

西部的院子是进入场馆时人流最多的"通道"。事实上，在原厂房中，它是中部院子向西的延伸，是改造中借用跨越南北车间的天桥界定出的一个特定的院子。此院的意义是：中院的前院，登堂入室的前奏，类似中国古典园林。因此，其本身就具有场所意义。为进一步强化场所感，设计中将拆出的天车梁成排错置埋于地下，形成界墙，再结合院中的树，在导引人流的同时进行围合。

Western Yard

The western yard is the passage of the most people on the way to the building. In fact, it was the extension of the central yard to the west in old factory, also it is a fixed yard by the flyover crossing the southern and northern workshops in transformation. The message of this yard is that it's the front yard of the central one, and the forepart of the following places is just like a Chinese classical garden. In this sense, it has the meaning of space itself. To enhance such meaning, we buried the breaking overhead travelling crane with certain shape and built into the walls, at the same time, guided the people with the help of the trees.

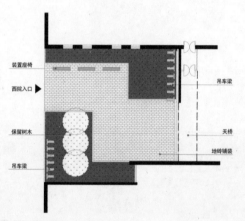

西部院子解析图（苍雁飞 绘）
The Structure Diagram of West Yard (Drawn by Cang Yanfei)

西部院子入口（李鹏 摄）
Entrance to the Western Yard (Taken by Li Peng)

西部院子入口局部
（张广源 摄）
Part of the Western Yard Entrance
(Taken by Zhang Guangyuan)

西立面及入口（莫日根 摄）
West Vertical Section and the Entrance(Taken by Morigen)

1.2 入口

建筑入口是室内空间动线的起点,也是室内外联系的中介。在严寒地区,入口还承担着防寒的特殊功能。同时,基于如下考虑,本案对入口给予了格外关注:

1.入口占据中部院子的核心区,是院子中唯一新置入的构体,对营造新旧交融的场景十分关键。方案选择了玻璃体,在强调通透的前提下表达了对既有场所元素和空间尺度的尊重。

2.扩大门斗,演变成一个可以驻足的空间节点。由此,该门厅在满足人流通过的同时还具有了如下作用:仪式接待、信息布告、等候约会、休息交流。

3.透过玻璃门厅看院子,院子中所有旧的景观元素都变成了"画"。这样的体验是置身院中时所感受不到的。

1.2 Entrance

Entrance to a building is the start of the interior circulation as well as connection of indoor and outdoor. In cold regions, entrance also protects against the cold. This design gives more weight to the role of entrance with consideration of the following reasons.

First, entrance takes the central and only new position in the yard, which emphasizes the combined scene of the old and the new. This design uses glass to ensure the transparency and meanwhile delivers the respects to both the site items and special scale of the given place.

Second, enlarging the foyer to make it a space node for stop. For this reason, besides the function of ensuring the people entrancing, entrance plays the other functions: ceremony reception, information notice, appointment waiting and for rest and communication.

Third, seeing the yard through the glass entrance, everything in the yard becomes the things in the picture. This experience is quite different from that when it is only located in the yard.

入口门厅解析图(苍雁飞 绘)
The Structure Diagram of the Hall Close to Entrace
(Drawn by Cang Yanfei)

入口、门厅与水池(莫日根 摄)
Entrance, Entrance Hall and the Pool(Taken by Morigen)

入口门厅南侧（方振宁 摄）
South Side of the Entrance Hall (Taken by Fang Zhenning)

从入口连廊看中部院子（方振宁 摄）
Central Yard Seen Through the Entrance Hall's Corridors
(Taken by Fang Zhenning)

从室内看门厅阳光房（方振宁 摄）
Entrance Hall's Glass Room Seen Inside of the Hall(Taken by Fang Zhenning)

从南侧看门厅阳光房（方振宁 摄）
Entrance Hall's Glass Room Seen Through the South(Taken by Fang Zhenning)

门厅多功能阳光房内部（曹扬 摄）
Inner Part of the Multi-Functional Entrance Hall's Glass Room
(Taken by Cao Yang)

1.3 中厅

中厅作为建筑馆动线序列中的核心节点,组织着内部空间的各种流线。如果说,在门厅看建筑馆内部还很朦胧,那么进入中厅便会有一种开门见山的感觉。

建筑馆中的这个空间不但是交通的枢纽,同时也因为视线的汇集、空间的共享以及小领域的限定使它变成一个产生更多交流和故事的场所,在此,日常的活动呈现出多样化的效果。

1.3 The Hall

As the core node in the interior circulation, the hall organizes the streamlines in the internal space. Compared with the misty view through the entrance, it instantly clears up when you come to the hall.

Besides the function of transport hub, the hall is also a space for more communication and exchange where the visions be centered, space be shared and a smaller space be limited.

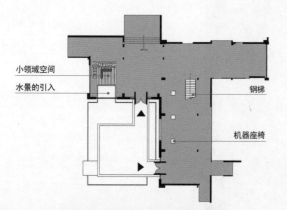

中厅解析图(苍雁飞 绘)
The Structure Diagram of the Hall (Drawn by Cang Yanfei)

中厅东侧(方振宁 摄)
East of the Central Hall (Taken by Fang Zhenning)

中厅的光影（张鹏举 摄）
Shadows in the Central Hall (Taken by Zhang Pengju)

中厅向西看（李鹏 摄）
Westward of the Central Hall (Taken by Li Peng)

1.4 梯

在建筑馆中，楼梯的设计除了满足竖向交通和人员疏散的功能外，另一个作用就是作为动线中产生更多交流的特殊节点。

在由"线"变"点"的动线设计中，楼梯及平台的作用是不可或缺的：进入教室区的直跑楼梯、平台联系着各层的教室，两向人流在此相遇；步入展区和评图区空间的台形楼梯，层层平台延伸出小型展区和交流场所。

此外，不同楼梯具有不同的性格：钢梯、直跑，传达着效率与快捷；砖梯、递进，散发着温馨和节奏。

1.4 Staircases

In a building, staircases are designed to offer more special space for communication in the whole circulation as well as to meet the need of evacuating the people vertically.

The staircase and platform are part and parcel for the design of "line-to-spot": straight run stairs link the classrooms together in different floors; the spiral stairway in the exhibit zone even extends the space for exhibition and exchange.

What's more, different staircases have distinct features: the steel stairs, for straight run, expressers the efficiency and convenience; the bricked stairs, is a sign of warmth and rhythm.

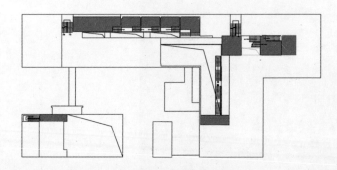

梯与平台的位置图（苍雁飞 绘）
The Location Plan of Staircases and Platform (Drawn by Cang Yanfei)

新旧楼连接处的梯子（张广源 摄）
Staircase Lying in the Connection Point of the New and Old Building
(Taken by Zhang Guangyuan)

中厅东侧的钢梯（李鹏 摄）
Steel Stairs in the East Side of the Central Hall (Taken by Li Peng)

中厅北侧的梯（孟一军 摄）
Staircases in the North Side of the Central Hall (Taken by Meng Yijun)

从三楼看中厅北侧的梯子
（孟一军 摄）
Staircases in the North Side
of the Central Hall
Seen from the Third Floor
(Taken by Meng Yijun)

1.5 "桥"

这里的"桥"是建筑馆中连接各功能区块的空中走道,它也是整个建筑空间动线节点中的生动环节。为了取得开敞空间的效果,并且能够让自然光尽可能多地引入建筑底层,空间通过"桥"的方式进行连接。同时,空间也被分隔成几个大小不一的垂拔,进而场所的艺术效果得到了强化,而"桥"本身也成为师生愿意逗留的节点。

"桥"因开敞空间中的联系而生,使动线在不同的区块间得以流通。同时,在这里,"桥"本身即有沟通、交流之寓意。

1.5 "Bridge"

"Bridge" here refers to space paths linking the other different function zones in the building, which is also the dynamic part in the whole circulation. To achieve the effects of opening space and letting more the natural light into the building, the spaces are linked together with "bridges". Meanwhile, the space is divided into different atrium spaces, by which the art effects are reinforced and the bridge also becomes a space interesting the stops of the teachers and students.

The bridges are designed to connect the open space and circulate the lines in the building. Besides it conveys the significance of exchange and communication among people.

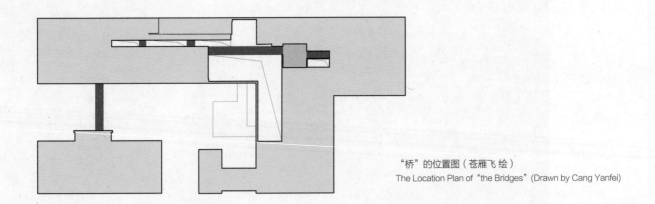

"桥"的位置图(苍雁飞 绘)
The Location Plan of "the Bridges" (Drawn by Cang Yanfei)

连接主体与物理实验室的桥(莫日根 摄)
The Bridge Connecting the Main Part and the Physics Laboratory (Taken by Morigen)

连接主体与物理实验室的桥（张鹏举 摄）
The Bridge Connecting the Main Part and the Physics Laboratory(Taken by Zhang Pengju)

从西部看跨越中厅东西的桥（孟一军 摄）
The Bridge Crossing the East and West of Central Hall,
Taken from the West Side(Taken by Meng Yijun)

从东侧看跨越中厅东西的桥（孟一军 摄）
The Bridge Crossing the East and West of Central Hall,
Taken from the East Side (Taken by Meng Yijun)

从中厅二层西望（孟一军 摄）
Watch Westward from the Second Floor in the Central Hall (Taken by Meng Yijun)

从一层西侧看跨越中厅东西的桥与公共空间（曹扬 摄）
The Bridge Crossing the East and West of Central Hall and the Public Space, Seen from the West Side of the First Floor (Taken by Cao Yang)

2 新场所的工业气质

每一个时代，都有其特定的、代表性的产业类建筑，其建筑本身的样式、风格及空间形态等都对它周边的环境施加着影响，并随着时间的积累增长了某种人文价值。

对于建筑馆，它的"原真性"保持良好，且之前未进行过功能转变和改建，因而，改造设计的各个环节都试图保持和纯化其特有的工业气质。在此，材料、装置、色彩、构造等的处理和运用都是十分关键的。

2 Industrial Disposition of the New Place

Each stage has its own unique and representative industrial architecture. The styles, patters and spatial form generate the influences on their surroundings and also develop its human and cultural values over time.

As for this building, since the originality is well kept without rebuilding and transforming any functions, each section of the design is trying to keep and purify its unique industrial disposition. For this reason, it becomes essential to decide on the use of the materials, equipments, colors and the structure.

南立面局部（方振宁 摄）
Part of the South Vertical Section(Taken by Fang Zhenning)

从门厅看烟囱（方振宁 摄）
Chimney Seen from the Entrance Hall
(Taken by Fang Zhenning)

南立面局部（方振宁 摄）
Part of the South Vertical Section(Taken by Fang Zhenning)

2.1 材料

为使改造后的建筑馆成为"源于工业时代"又"与旧时期不同"的建筑,建筑材料的运用至关重要。

材料的肌理如同皮肤,其质感传达着某种表情。建筑馆材料的选择主要是改造再利用原有厂房中的废弃材料,加入少量新的材料,使原建筑携带的人文记忆得以延续,同时,在新功能转换的过程中,在新与旧之间创造一种对话,使旧有的形态更加凸显。

钢构件的大量选用突出反映了建筑原有的工业气质;黏土砖的反复出现又在冷峻的工业气质中增添了温馨的怀旧情调;简单的水泥地面使得教学场所显得粗犷而直接;通过局部U玻、单玻等光洁材料的衬托,整体试图传达出工业建筑特有的清晰、朴素、单纯和率真。

2.1 Materials

In order to reconstruct this museum into a building originating but different from industrial age, it becomes essential to decide which construction materials to use.

The mechanism of a material is like the skin, expresses some kind of feeling. To reserve the humanistic value, this museum mainly used the waste materials of the factory buildings, together with a small part of new materials, which also ensures a transmission between the old and the new forms on the basis of highlight of the old form.

A considerable use of steel materials highlighted the industrial disposition of the former building. The large number of clay bricks added more reminiscence emotions to the cold and solemn industrial disposition. The simple cement flooring makes this teaching place more vigorous and unconstrained. With using some bright and clean materials serving a foil, such as the U-shaped glass and single glass, the whole building manages to express the typical dispositions of industrial buildings, including clearness, plainness, simplicity and directness.

建筑馆的材料(张鹏举 摄)
The Materials of the Architectural Hall(Taken by Zhang Pengju)

从三层东望中厅（曹扬 摄）
The Central Hall Seen from the Westward on the Third Floor (Taken by Cao Yang)

2.2 装置

旧产业建筑作为那个时代工业文明的历史见证,有着辉煌的昨天。这些工业建筑伴随着人们的生产生活,早已写满了他们的记忆,留在原厂房中的种种装置也是那个时代的烙印,对它们的改造再利用,会让人们产生更多的联想。

原厂房中的机器装置被完好保留,或改造成特殊的构件,或成为校园中的"艺术品",它们像一个个伫立着的"老人",冷静地凝视着新旧时代的交替。它们面向着今天而又强调着历史,例如艺术沙龙与办公区中的冲天炉、模型室中的大型"沙漏"、庭院中的煅烧炉、草坪中的机械装置以及整个建筑馆中交织着的各式管道等。

2.2 Equipments

The former building which witnessed the industrial civilization in that age had a glorious history. These industrial buildings have accompanied people's lives and impressed up their memories fully. The equipments left in the factory are carrying the stamp of that age. It helps arouse more memory and imagination of people to reconstruct and make use of them.

In this project, all the equipments from the former factory are well-kept, some of which are transformed into devices for special use and some other into the works of art placed on the campus who are staring at the changing over generations like some elderly men. Whatever the cupola located between the art salon and office area, the huge sandglass model in the model room, the calcinatory in the yard, the mechanical devices on the lawn and interweaved pipes in the whole building, they are all looking into this present age and also highlighting the history to us.

建筑馆的装置(莫日根 摄)
The Equipments of the Architectural Hall(Taken by Morigen)

沙龙局部（曹扬 摄）
Part of the Salon (Taken by Cao Yang)

2.3 色彩

色彩是一种语言,能触动人们的情感。不同时代的人们对色彩有着不同的审美反映,同时,不同时代的建筑场景又有着不同的色彩集合。原厂房的整体建筑颜色来自于那个年代的材料及建造技术,正是这一特定的色彩集合给予了新场所中人们更多的联想。

对于建筑馆,砖、木、水泥的固有色是曾经年代的建筑色彩集合,质朴而有温度,加入钢构件的冷灰色调,进一步染浓了曾经工业时代的气质。冷暖交替而相融,暖是灰色的暖,冷是灰色的冷,这正是新场所特定工业气质的魅力所在。

2.3 Color

Color is a kind of language, which stirs up people's feelings. People in different ages have different appreciation of the colors. Similarly, architectures in different times prefer different uses of colors. The color set the former factory adopted was determined by the material and construction technology in that age. It is this color set that is capable of inspiring the people placing themselves in the new building.

For this building, the colors of the bricks, woods and cements preserved the former color set. Meanwhile, the cold grey color of steel structures adds more emphasis on the ever industrial disposition. With the use of cold grey and warm grey, the mixture of cold and warm colors represents the particular industrial disposition of this new building.

建筑馆温暖怀旧的色彩
(莫日根 摄)
Warm and Nostalgic Hue of the Architectural Hall
(Taken by Morigen)

丰富色彩的中部院子
（孟一军 摄）
Colorful Central Yard
(Taken by Meng Yijun)

2.4 构造

建筑馆的改造重视旧厂房的主体结构，不对其进行随意或过分的拆改，仅做些必要的、小限度的改动，使其有效地转化为新的使用功能，这是保留工业气质最根本的策略。

在改造过程中，技术方面首先面临一个结构加固的问题，加固的方法是直接在最薄弱的部位使用型钢强化；功能方面则面临一个加层的问题，加层的方法是直接在钢柱梁之间浇筑混凝土楼板；在其他构造方面，采用新旧并置并直接连接碰撞的策略，如门、楼梯等。所有的这些做法都将构造裸露，简单有效，并都强调一种直接的制造感和一种工业的率真感。

2.4 Structure

During the reconstruction process, to transform it into new functions, the main structure of the former factory was respected without some unplanned or too much removal. Only some necessary and changes on a small scale are made. These measures prove to be the most significant strategies in preserving the industrial disposition of the building.

During the process, a technological challenge we face is how to fasten the structure. The measure we adopted is to strengthen the weakest part with the section steel. As for the function, the challenge is story-adding. The measure adopted is to place the concrete floor boards between the steel beam-columns. In other aspects, the new building gives weight to both the juxtaposition and collision between the old and new structures, such as the doors and stairs. All these measures exposed the internal structure, simple but efficient. What's more, it highlights the sense of manufacturing and the directness of industry.

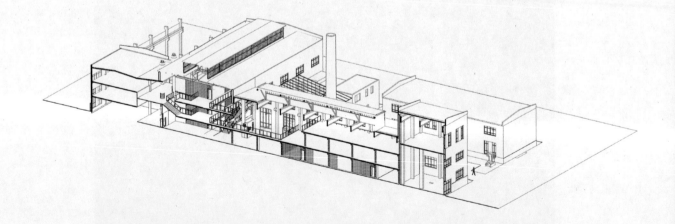

建筑馆剖视图（薛飞 绘）
The Cutaway View of the Architectural Hall(Drawn by Xue Fei)

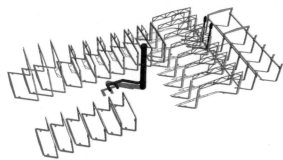

原有结构（苍雁飞 绘）
Original Structure(Drawn by Cang Yanfei)

结合加层的结构加固（苍雁飞 绘）
The Reinforcement Structure Combined with Storey-Adding(Drawn by Cang Yanfei)

建筑馆的构造（扎拉根白尔 摄）
The Structure of the Architectural Hall(Taken by Zhalagenbaier)

3 新场所的故事场景

现实生活中的事件都有其发生的特定"场所",创造了事件的发生地也就创造了"故事"。而随着故事的上演,场所也由客观的、具体的事件发生的转变为带有主观色彩的"场景"。

创造"场景"有许多方法,一般地,须先创造空间的领域性,而领域性的获得通常又有两种方式,一是围合,二是向心,前者须有"界面",后者须有"心点"。这些"面"与"点"在建筑馆的改造中常常被设定为刻写了曾经"事件"的"构件"和"设备",或言,倚借这些留有历史印痕、带有触感的构件及设备,"场景"更容易发生。

3.1 烟囱下的对话

建筑馆中的几处烟囱是场所极具个性的视觉元素。

入口处的砖砌烟囱是校园中识别建筑馆的重要特征。它位于院子中央,散发着视觉张力。作为倚借物,师生们愿意在此驻足交流。为强化它的作用,设计有意将其组织为进入门厅的必经之地,并与雨棚相结合。雨棚之于烟囱正如伞与柄,二者的有机结合使其下的对话成为一种经常性的行为。另一处烟囱位于东部办公区,由沙龙的两个冲天炉向上延伸而成,同样的作用在设计中予以再次强调。

3 Story Scenes in the New Place

All the events in real life have their specific places where they occur. When a place is created, the story is too. As the story unfolding, The place also transformed from an objective and specific event into a "scene" with subjective feelings.

Many ways help create scene. Generally and firstly, create a territoriality of the space. Congagement and centripetal are two ways to achieve the territoriality. The former requires a surface and the latter a spot. These surfaces and spots often play the role of units and components with some even stories printed on. In other words, scenes are more easily created with these units and components carrying some historic prints.

3.1 Dialogue under the Chimney

A couple of chimneys in the museum become the very specialized visual elements for this building.

Among them, the brick chimney at the entrance, situated in the centre of the yard, emanating visual forces, becomes the important sign to help identify this museum. It also offers as a place for resting on and exchange of the teachers and students. To reinforce its function, this design put it together with the canopy and made it the only access to the hall. The canopy for the chimney is like the umbrella and its handle. The combination of these two makes it possible for the frequent dialogues and exchanges under the chimney to take place. Anther chimney is located in the eastern office area, consisting of two cupolas, giving another weight to the same value.

3.2 熔炉前的聚会

在中部院子,保留了一个煅烧炉,它是拆出部分维护砖后自然形成的院中院,它的形态似乎印刻着曾经年代生活和劳动的剪影,吸引着过往的师生,加之围合感极强,在室外空间中形成一个领域性很好的私密场所。因此,熔炉前的聚会自然成为一种常态化的场景。另外,对于沙龙的两个冲天炉,在其间加设了木地板,自然划定了领域,其上可表演可茶歇,都是抢手的场所。

3.2 Meetings in Front of the Fireplace

A calcinatory is reserved in the centre of the yard. It is a yard in yard formed after removal of some bricks. It interests the passing stuff by reflecting the past life and work. Together with the strong sense of congagement, it becomes a private space in the outer space. Hence, it becomes a frequent scene to meet before the fireplace. Besides, wooden floors are added between the two cupolas to form a natural space which can be used both for performance and tea time, undoubtedly, become popular space among the stuff.

3.3 漏斗旁的互动

建筑馆西侧模型室顶部成排的沙漏形式感极强,由此可联想到一种漏斗旁互动制作的场景。反过来,由此衍生出的模型制作功能则又重演了这一场景。对这个空间而言无需做任何改动,清理卫生是唯一的工作。

3.3 Interaction beside the Funnel

In the west of the Architecture Hall, on the top of the model room, the sandglasses in rows generate a strong scene of form, which arouse the imaginations of a scene of interaction beside the funnel. On the contrary, the derived function of making models replays this scene. It's no necessity to make any change to this space, except for ensuring the garbage cleaning.

3.4 梁架下的交流

建筑馆东部院中裸露的梁架是另一处具有个性的特殊构件，它记载着原厂房自身生长的痕迹，具备一种极有趣味的围合感。因而，对于场景的创造，本身并不用去加建什么，抬起地面即成"舞台"，同时也成为室外"茶吧"。同样，室内也保留了裸露的梁柱和各式的管道"构架"，它们在顶部及周边硬质接口下对空间进行了二次界定。这些"构架"所具备的近人尺度和朦胧感使其下的交流场景更容易上演。

3.4 Talking under the Beam Frame

Located in the yard in the east of the Architecture Hall, the exposed beam frame serves as another specialized component. It carries memory of the former factory's development and history, with an extreme sense of congagement. With no need to add anything but to set up the ground, it forms a stage as well as a tea bar outdoors. Similarly, beam frames and various pipes are also reserved indoors. These frames and pipes actually divide the space under them for the second time, which makes it possible for the exchange under them with their neighboring and misty features.

张鹏举 摄
Taken by Zhang Pengju

4 新场所的生态机能

无需多言，厂房的改造理念首先基于生态。然而，保留和节约并不意味着生态的全部，从人本和发展的角度看，改造后新场所的生态效应同样需要重视。因此，在改造过程中，赋予场所更加完善的生态机能、挖掘适宜的生态策略同样重要。

4 Ecological Mechanism of the New Place

Undoubtedly, the reconstruction of the factory is firstly based on the philosophy of ecology. Preserving and saving are not all for the ecology. From the standpoints of human and development, the ecological effects after the reconstruction deserve more attention and respect. Hence, during the reconstruction, it becomes the same essential to generate more complete ecological functions and strategies.

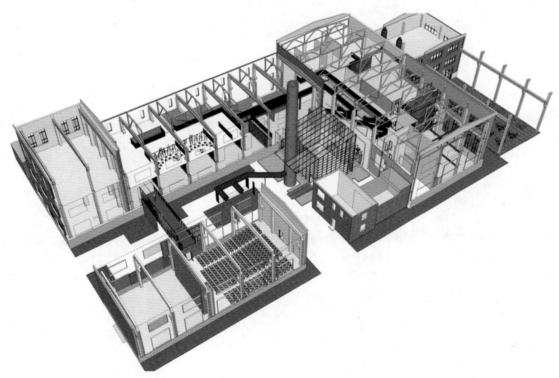

充分挖掘新场所的生态机能（苍雁飞 绘）
Fully Devlop the Ecological Function of the New Location(Drawn by Cang Yanfei)

4.1 光

光在建筑馆中是最微妙、最具精神性的元素。在满足采光要求方面，天窗作用的充分发挥解决了加层后空间深处光线不足的问题。在此基础上，多处设置天井，使得导入的阳光创造了丰富的光影效果，同时，使用的多种玻璃散射出朦胧的光线。总体而言，在温暖的砖墙、水泥和冷峻的钢构面上，丰富的光影和弥散的光晕使场所具有了某种艺术氛围。

4.1 Light

Light is the most delicate profound component in the Architecture Hall. To meet the lighting demands, use of skylights supplements the lack of light in the deep space caused by story-adding. Based on this, several open yards create various lighting effects. Meanwhile, different kinds of glasses generate misty lights. In general, on the warm brick walls, cement surface and cold steel surface, abundant light and light halo have built the art atmosphere.

空间的光与影（张鹏举 摄）
The Spacial Light and Shadow
(Taken by Zhang Pengju)

红砖梯上的光影（张鹏举 摄）
Shadows on the Red-Brick Staircases
(Taken by Zhang Pengju)

4.2 风

就创造风环境而言，利用已有设施以及高大开放的空间组织被动式室内气流是本项设计的主要创新点之一。具体而言，旧厂房的天窗、烟囱、地道等组成部分构成了潜在的自然通风路径，巧妙地对其加以组织不但能够实现通风还可以降温、加湿，并保证了室内空气的质量和环境舒适度。在改造中，室外的新鲜空气经地下通道冷却后进入室内，再经可电控的天窗向上提升，组织了开放空间的自然通风，同时，通道入口处设置的水池将湿气带入室内；烟囱则组织了报告厅、计算机房等封闭房间的通风。

4.2 Ventilation

As for the ventilating environment, with the help of the existing facilities and the high and open space, this building innovated in generating a kind of forced internal airflow. To be specific, the skylights, chimneys and tunnels of the former factory have formed the potential natural Ventilation paths. Proper reorganization of these not only ventilates but cools, wets and refreshes the air, ensuring the conformability of the indoor air. During the reconstruction, the fresh air comes indoors through the tunnels after cooling and rises throng the skylights. This generates the natural ventilation of the open space. At the same time, the pool at the entrance to tunnels carries the wet air indoors. Chimneys ensure the ventilation in the lecture hall and computer rooms and other closed spaces.

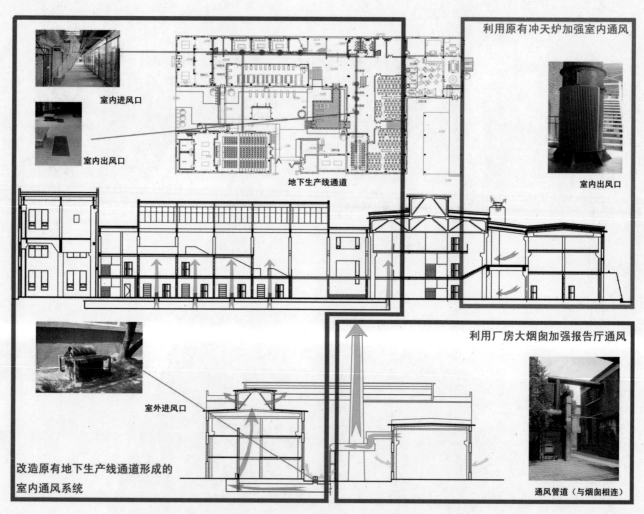

薛剑 绘
Drawn by Xue Jian

4.3 绿

建筑馆植物方面的生态策略体现在保留周边有价值的树木。东边院子中的许多树木以及入口处的大树都在设计和施工中做了精心保护，如路径形式和门厅形状的设计均来自于对树木的退让，因而，改造设计在尊重和创造环境的同时也构成了"绿"景。

4.3 Green

The ecological strategies on plants are embodied in some valuable trees around the building. Trees in the eastern yard and at the entrance have been carefully protected in the reconstruction. The design of the path and hall proved this. Therefore, the reconstruction is respecting and recreating environment as well as creating a green scene.

建筑馆中保留的树木（张鹏举 摄）
Trees Reserved in the Architectural Hall(Taken by Zhang Pengju)

建筑馆与保留树木共生（苍雁飞 绘）
The Building Coexists with the Reserved Trees(Drawn by Cang Yanfei)

树荫下静谧的东侧院子（张鹏举 摄）
Peaceful East Yard under the Shade(Taken by Zhang Pengju)

秋天的植物与建筑（莫日根 摄）
The Plants and Buildings in Autumn(Taken by Morigen)

与植物交融的南侧入口局部（扎拉根白尔 摄）
Part of the South Entrance Mingled with Plants (Taken by Zhalagenbaier)

4.4 材

建筑馆用材方面的生态策略主要表现在废材利用和可再生材料的使用上。对于前者，主要表现在厂房废旧材料的利用和构件转换功用易位再用，如，废砖用于铺地；钢窗用作栏杆；吊车梁变成支柱；旧机器转换为艺术装置；旧有的传输带平移后成为过桥；裸露的钢筋混凝土牛腿成为上层结构梁的支托；拆出来的零星废旧钢板拼成图案用于门窗洞的合理封堵；同时，大量的天车导轨和传输带的组合构件变成了钢梁、钢柱、钢梯的主材。

在可再生材料的使用方面，主要表现在使用钢材和玻璃上。同时，其快速安装的特点，为建筑赢得了更长的使用时间，这也是绿色生态思想的一种具体体现。

4.4 Materials

The ecological strategies on materials are embodied in reuse of the waste stuff and use of renewable materials. The waste stuff reused include the waste materials in the factory and some devices, such as the waste bricks used for paving the floor, steel windows for railing, beams of cranes for pillars, old machines transformed into works of art, old transmission belt becoming the bridge, exposed reinforced concrete into the supporting bracket to the top beam, scattered steel boards into some patterns to block the holes in the doors and windows, and also combined parts of crane rail and transmission belt into the main stuff for steel beams, steel pillars and steel ladders.

On the aspect of choosing the renewable materials, it mainly uses the steel and glass. At the same time, it can be installed quickly so it can save time for architecture, also it is a kind of concrete embodiment of green ecology idea.

废旧材料利用组图（薛剑 绘）
Pictures of Using Waste Materials(Drawn by Xue Jian)

东侧的院子，由废钢窗、废旧砖做成的围墙（扎拉根白尔 摄）
East Yard, Enclosing Wall Made of Waste Steel Windows and Bricks(Taken by Zhalagenbaier)

生长 从纯粹走向复合
ENLARGEMENT From Simple to Compound

内蒙古工业大学建筑馆改造完成后，在其北侧扩建一栋用于建筑类专业教学的设计楼，由专业教室组成，规模7000平方米。在分析了功能和基地现状的基础上，设计的对策集中在从老馆到新馆的有机生长方面。这是一个健康的建筑所必然应该经历的过程。对于本项目，突出表现在以下方面：动线的过渡、空间的同构、机能的延续、形式的生长。

When the conversion of the Architecture Hall of IMUT has been finished, a designing building which covers 7000 square meters, made up by the professional classrooms, will be built for building professionals on the north side. On the basis analysis of the function and general base situation, the strategies should focus on the organic growth from the old building to the new one. This is the process a healthy architecture has to experience. This project is highlighted in the following aspects: the transition of fixed line, the isomorphism of space, the continuing of function and the growing of form.

扩建设计楼西立面（陈溯 摄）
West Vertical Section of the Expanding Designing Building (Taken by Chen su)

薛飞 绘
Drawn by Xue Fei

设计楼与旧楼之间的内院（陈溯 摄）
The Inner Yard of the Designing Building and the Old Building(Taken by Chen su)

1 动线的过渡

设计伊始，选择与老馆的连接方式成为思考的切入点，这是一个新旧馆之间动线过渡的问题。场地条件和建筑功能决定了新建筑形体的走向，而老馆适宜的连接位置早已在改造时预留。因而，设计的对策是在二者之间增设一个门厅。此门厅既是二者空间的过渡，又是从北部直接进入新馆的入口，同时更成为师生们新的交流场所。为进一步加强动线的流畅感，沿此厅进入新馆后，有意增设了直跑楼梯用于引导人流。

1 Transition of Fixed Line

At the beginning of the program, the entry point of designing is to choose the way to connect the old building and new one. This is the issue of transition of fixed line. The conditions of the site and the function of the architect determine the alignment of the shape of the building, While the proper position for the connection is preserved. Therefore, the strategy is to build an additional hallway between them. This hallway is the transition between the old building and the new one and it is also the entrance from which the north section to get into the new building. At the same time it provides a place for the teachers and students to communicate with each other. In order to increase the fluency of the fixed lines, after entering the new museum along with the hallway, a straight run staircase is designed to lead the flow of people.

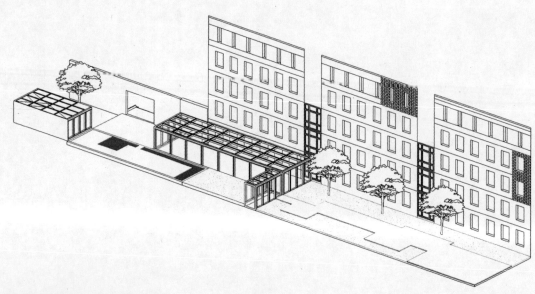

新旧楼之间内院界面构成图（薛飞 绘）
The Space Diagram of the Inner Yard Between the Old Building and the New Building (Drawn by Xue Fei)

从西北向看设计楼(张广源 摄)
The Designing Building Seen from the West-North (Taken by Zhang Guangyuan)

新建楼东侧入口局部（陈溯 摄）
Part of the East Entrance of the Newly-Built Building (Taken by Chen Su)

新建设计楼与旧楼连接处入口局部（张广源 摄）
Part of the Connecting Entrance of the New Designing Building and the Old Building
(Taken by Zhang Guangyuan)

2 空间的同构

以走道串联功能单元成为新空间布局的基本模式,这是一种强调效率的普遍形式。设计在此基础上,继续传承旧馆的质量,在空间组织上与旧馆具有同构性。如,围合入口处的室外空间再成院落,在新馆的动线中重新增加了具有园林质量的新起点;扩大走道和单元间的空间,使联系空间再次具备了适宜交流的场所性,并在特定的时候成为功能的主体。

2 Isomorphism of Space

The basic model to connect functional unit through aisles is a common way to emphasize efficiency. On the foundation of this, this design inherits the qualities of the old building. In the organization of the space, the old and the new building, have the isomorphism of space. For instance, the space outside of the entrance of the enclosure place is another yard; a new starting point with feature of garden is added, the aisle and the inter-unit space are expanded, so that the inter-space is provided with the communicational function which becomes the major function of the hallway.

新旧楼院子东望(陈溯 摄)
Watch Eastward from the New and the Old Building (Taken by Chen Su)

设计楼西侧连廊(李鹏 摄)
Corridor in the West Side of the Designing Building (Taken by Li Peng)

3 机能的延续

新馆继续注重建筑的自然生态性，在设计上集中表现在功能单元间的横向分离，由此，建筑获得了以下生态机能：风——在各层相对分隔的前提下，在分离处的空间上方每层设独立的竖向通风井，利用热压组织自然通风；光——利用分离后南北向的间隙引入自然光，使较深的体量内部有了理想的光环境；绿——同样的体量分离，使南北侧保留的大树引入室内公共空间，建筑和环境再次相融。

3 Continuity of Function

The new building emphasizes the natural ecological function of the architect, which is reflected in the horizontal division of the functional unit. Therefore, the building gains the following functions: wind—on the condition of every floor is relatively separated, above the division spot vertical shaft is set up separately in every floor. Hot organization is adopted for natural ventilation. Light—natural light is led in through the gap between north and south after separation, so that the inside part of the building will also be well-lighted. Green—the same separation makes the reserved trees in north and south sides enter the public space—so that the man-made environment and the natural environment integrate with each other harmoniously.

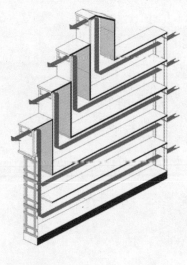

薛飞 绘
Drawn by Xue Fei

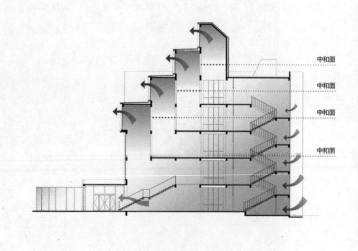

薛飞 绘
Drawn by Xue Fei

4 形式的生长

在形式上，新馆放弃延续旧馆的形式语言，力求简明，并期望表现新时代的技术和功能特征，以一个旁观者的身份注视着原有建筑，表达一种也许更为深刻的尊重，同时也展示了校园环境生长的历史过程。在此基础上，新馆又用清理场地拆除的废砖作为"符号"，砌筑部分墙体，传达了新旧馆之间共同的生长基因。

4 Growth of Form

On form, the new building abandons the formal language of the old building, and tries to pursuit simplicity to demonstrate the new age's features in technique and function. It watches the old building as an on-looker, to express the deep respect. At the same time it present the historical process of the developing of the campus. On the foundation of that, parts of the wall of the symbolic building in the new building are built by the waste bricks that had been got from the removing the space which conveys the common genes shared by the old building and the new one.

设计楼内单元分离后的竖向中厅
（张广源 摄）
Vertical Central Hall Separated by Different Sections Inside of the Designing Building
(Taken by Zhang Guangyuan)

设计楼中厅向上仰视(张广源 摄)
Looking Up in the Central Hall of the Designing Building (Taken by Zhang Guangyuan)

设计楼中厅立面局部（张广源 摄）
Part of the Vertical Section in the Central Hall of the Designing Building
(Taken by Zhang Guangyuan)

设计楼外立面局部（张广源 摄）
Part of the Vertical Section Outside of the Designing Building (Taken by Zhang Guangyuan)

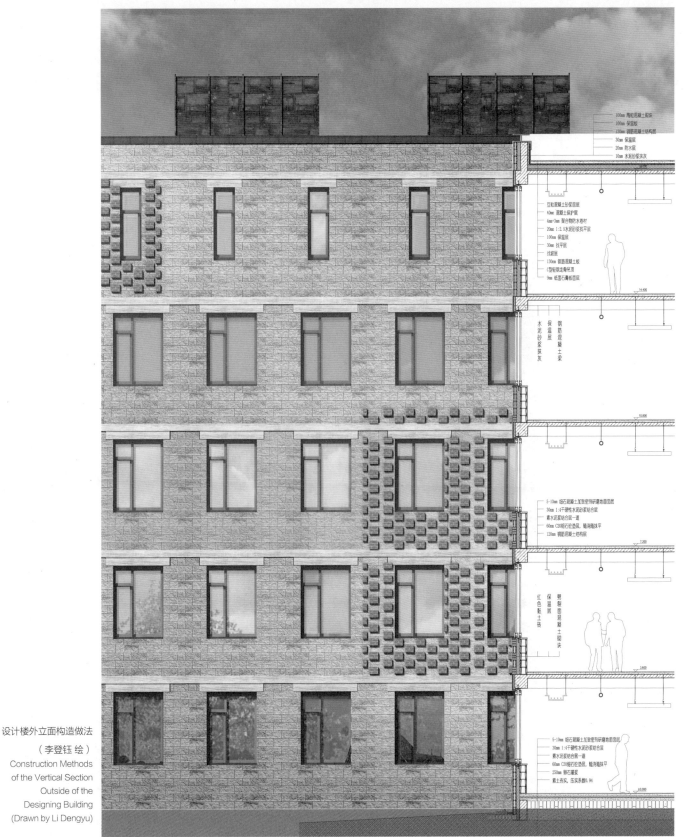

设计楼外立面构造做法
（李登钰 绘）
Construction Methods of the Vertical Section Outside of the Designing Building (Drawn by Li Dengyu)

从东南向看设计楼（张广源 摄）
The Designing Building Seen from the East-South
(Taken by Zhang Guangyuan)

从西北向看设计楼局部
（莫日根 摄）
Part of the Designing Building
Seen from the West-North
(Taken by Morigen)

二、关于建筑馆的思考
II. Thought of the Architectural Hall

从校园里的旧厂房开始

本人作为一名在大学里教书的从业建筑师，常被邀参与关于校园建设的讨论。本年初，在同样的一次讨论会上，学校准备启动拆除校园内一组旧厂房的计划。当把这一消息带回到工作室，同学们的情绪沸腾了，随即开始了以旧厂房的去留为主题的讨论与研究。而这次参会也启动了我为此工作了近一年的一项设计实践，同时也启发了我对一些"平常"问题的关注和思考……

一、关于旧厂房的存留

"简单的机器和寂静的厂房身上充满着人们对过去的感受和怀旧的思绪，伟大的工业时代已经结束了的感觉，使这些静物给人带来忧郁的感受……"

——历史学家哈伯特·哈比森

通过对这一组厂房建筑的现场调研和技术咨询，这一环节讨论的结果是，向学校领导上呈《关于旧厂房保留改造的建议书》，申请通过改造赋予它新的生命，其中主要陈述以下三个理由：

1.这批遗留建筑是学校发展的记忆载体。校办工厂建立于20世纪50~70年代，对当时学校的"产学研"结合发展曾起过重要作用，90年代至今，随着产业结构的转型和办学方向的转变，这批厂房渐渐成为了历史，但它们记录着一个学校的发展历程。

2.废弃的厂房经过改造和设计，将成为校园独有的景观，具有特殊的艺术价值。这些厂房和设备是刚刚过去的那个时代的记载：厂区特有的秩序，厂房独具的空间特征，设备和机器的工业性技术美的造型……这些只需稍加处理就可传达出独特的场所感。旧砖的红暖色与现在冷灰色的新建筑相互映衬，使校园景观具有独特的艺术感染力。

3.废弃的厂房经过改造即可获得再生，产生适宜的使用价值。"厂房中唯一奢华的就是空间，在这里的生活是全新的理念，空间中有室内景观，设计师的意图使它充满活力，就像我们的城市一样，保持了空间开敞。"我校现有的厂房也同样具备这些特质，它们经过结构加固和设施改善，即可成为大学中重要的活跃元素。

出于某种说服力，报告中"不拆再用"的诸理由均冠以"生态"；总结之，一说文化生态——基于校园生长的历史记忆，再说资源生态——基于建筑全生命周期的观点。

当下，建筑设计常被某种时尚理念引领着。建筑师也许该常常关注一下身边那些朴素的房子，关注一下它们那些朴素的美德。校园中的这一组房子，位处我每日上课的必经之地，对它熟视无睹了20年，当被猛然关注之后，发现它竟显现出被岁月掩盖了的众多质量：真实的形态空间，真实的材料结构，以及由此产生的朴质、简约，进而儒雅。

二、关于存留旧厂房的用途

大学原本意义是——思想的市场，人们从那里能够极力寻求各种有意义的思想。

——《建筑模式语言》

这一议题主要分几个层次进行讨论：

1.当前校园里最缺乏的场所空间

保留厂房做什么？同学们的回答是惊人的一致——我们最需要的是校园文化活动空间。校园现状是由教学楼、食堂和宿舍组成的标准大学教育生产线，循环的工序周而复始，又加之工科院校的课堂上鲜有思想交流，同学们对校园文化生活的渴求是不难理解的。可是如果把这一点上升到理论高度，也还需要进一步的探究。

2.建立大学校园文化生活的目的与意义

（1）关于大学教育的内涵和基本目标

古代南欧初设大学的目的是为了培育学生从需要监护人的未成年人转变为有独立思考判断能力的成年人。教育之重点应该是以人为本位，教育目标是人的整体智能与道德的提升而不是单纯的知识或技术内容。现代大学教育的基本目标应该是：培养具有理性精神、独立人格，具备人文素养的"可通用的专业人才"——"通才"应该具备：理性精神、独立人格、人文素养、专业素质。

（2）大学原本意义是——思想的市场
人们从那里能够极力寻求各种有意义的思想。外部环境必须提供一种活动场所去鼓励而不是妨碍个性的发展和思想的自由，周围环境必须提供一种活动场所去鼓励学生理解、辨别哪些思想是有意义的；并向他们提供非常充分的机会去接触各种不同的思想，以便他们自己做出决断。

（3）情商（交流和表达能力）的重要性
美国最著名的成人教育家戴尔·卡耐基在多年给工程技术人员上课时发现，收入最丰厚的不是对工程学懂得最多的人，而是一个拥有专门知识，加上能够表达他的意念，并善于为人处世，领导和鼓舞他人的人。

（4）校园文化生活是促进自觉交流和表达的唯一有效途径
在成功的校园文化生活中，青年学生通过交流和表达来认识自我、促进自我教育，从而成就自我。就技术教育来说，当代各学科之间分工愈细，各自愈深的研究，导致学科间的合作和共同创作的可能性愈小，而其必要性存在强烈的潜在需求；因而必须关注相关学科的前沿发展（技术方面、艺术方面、哲学等方面）。其前提是从事各个学科的人们之间要有顺畅而积极的交往与交流。

于是，这一环节的讨论结果是：旧厂房应该改造为"学校小区文化活动中心"，其主要用途有——师生室内文体活动场地：体操、球类活动；休闲娱乐场地：茶室、书吧、沙龙、电影院；业余艺术学校，校园商业；展陈、表演……来支持教工和学生的课余生活，有了适宜的空间支持，交往与交流自然会产生。

有了上述充分的论证，我们向学校做了汇报。校长看着"未来的校园小区活动中心"的建筑动画和以"798"、"LOFT"的精彩照片展望的未来，突然问道："如果改造成建筑馆会不会比新建省钱？"

三、从旧厂房到建筑馆
建筑师的职责是为已经存在的便利提供空间，为尚不具备的便利提供可能性。

—— 路易斯·康

在本次设计实践中，当基于生态考虑时，大师的话也许要反过来说。

1.一次特别受益的毕业设计课程
我们心目中的校园小区活动中心将成为建筑馆，建筑学人将从此拥有属于自己的家，这是建筑系的师生一直以来的愿望，激动之余，面临的是要冷静处理摆在面前的问题。首先：我们的初衷是校园小区活动中心，接下来该怎么办？第二：厂房中奢华的空间如何控制利用？第三：如何解决资金不足的问题？此时正是毕业设计选题的时候，于是，就有了一次不同以往的毕业设计。

在毕业设计任务书中我们给出了上述问题和对于建筑馆的相关思考，在辅导同学们设计的过程中，问题的答案渐渐明朗地现出了：

（1）尊重场所感——开放式的建筑内外环境空间设计，营造建筑馆为精神意义上的校园小区活动中心。由于学校对专业建设的重视，建筑系偏得了这一具备特别场所感的宝地，但我们深知其原本的使命是校园文化活动中心，但这却正恰恰暗合了建筑学本属交叉学科的专业特点——这里的场所特质是平等、开放和愉悦的。

（2）生态为先——以最适功能原则、积极主动原则从建造现场出发综合多效地解决问题。这不仅仅因为有限的资金，更有主动认识的因素。乃至后来，建筑馆的功能已不再是思考的范畴，全部的努力都集中在寻找某种适当的生态策略；这些实为基于生态建筑学的人文精神，也正体现建筑师之创造性地解决问题的职业素养。

在指导毕业设计的同时，我得到了以上两项主导思路来指导下一步工作；本组的同学也经历了实际项目的历练，他们亲手测绘旧建筑，预算清理可重复利用的厂房构件，再运用自己的专业技能和修养，以绝对的积极和热情创作了母校的建筑馆，相信他们在以后的工作和学习中对设计会有不同的认识；作为教师的我也从同学的热情中感染到了创作的激动和教学相长的愉悦，并一直延续到现在正在进行的工程实践中……

2. 一次特别的生态工程实践
伴着毕业设计这项工程进入了实施阶段。
就厂房的利用而言，其最珍贵的质量莫过于空间。而在已有的空间和场所特质中安排功能，最恰当的办法或许莫过于看场所空间对功能的适应性，而不是像一般的创作过程那样为某个功能提供一个适当的空间，以及为一组空间组织动线和分区，这样的切入点为本次工程实践奠定了生态的基调。

（1）场所感引导功能。由场所决定功能可以最大限度地发挥空间的效率。当然，在为某个功能选择相应的空间时，我们不仅关注其物理尺寸，还关注其适宜的场所气氛。如，东南侧的独立车间，由梁架构成的秩序感和充足的光线是一个绝好的报告厅；西北侧的一处空间内，一些无法叫出名字的设备，按照特有的秩序摆放着，它们提供了可供操作和展示的平台，因而是一个极好的构造与材料陈列的空间。若再加上模型制作，设想学生们倚着这些旧设备，在不同的高度上参观和操作……这种场所氛围已超越了一般意义上的生态策略，一种特定的历史记忆融在其中。

（2）场所感引发功能。东侧的一处空间呈现一幅休闲的场面同时透出一种清风飕骨式的酷劲，于是在我们的任务书中加上了"咖啡厅"这三个字；还有一处用来烧铸的小空间，内部的烧陶炉依然可用，由此引发了一个新的功能——陶吧，也引出了一门选修课……在所有的生态策略中，最大限度地发挥空间的效率是首要的策略，同时，创造使身心愉悦的场所空间则是生态策略的终极目标。

（3）用记忆来营造新的场所空间。在清理环境中拆除附属房的旧砖正好成为新增的墙体材料和广场铺地；平移旧有的材料传输带演变成为一个传输师生的过桥，解决了西南部独立车间与主体的空中动线问题；同时，工厂原有的那种率真也带进了新的空间；而传输带的组合钢柱提供了所有新增楼层的钢柱用材；原有钢窗料改装为栏杆；大型钢筋混凝土吊车梁成为报告厅的排椅和环境中的座凳；裸露的钢筋混凝土牛腿则自然成为上层结构梁的最佳支托，而由此成全的迥然于现代层高尺度概念的顶层空间，反成了空间中的一处独特场所；旧有的机器置于室外环境，径直成为营造气氛、唤起记忆的当代艺术装置……

在本案的实践中，还采取了其他的生态策略，如被动的结构加固与功能加层的结合；排烟用的烟囱演变为通风用的装置；天窗在光和风之间找到最佳的平衡点。当然，程序做法的策略还有许多，但感触最深的是："生态"决非历经几个规定程序且更非口号，只有变成一种主动行为，并由衷倾注热情时，建筑才能真正"生态"。多么希望建筑师们都有一次类似的经历！所幸，我和同学们一起经历过了。

回想有关旧厂房的工作经历，在和学生们的讨论和同事们合作的过程中，我对建筑和营造有了与以往不同的体会……在厂房特定的场所空间中，在极为有限的资金投入下，来营造期望值很高的建筑，心情倒觉得比以往"创作"时更加从容、淡定；常常游走在工地上，不经意间又得到了一个妥善的解决方法时，有一种特别的轻松感觉……在这种可以得到意外褒奖的内驱力下工作；取舍之间、决断之处的无奈与欣喜、感怀与动容——使笔者切实感到了作为建筑师的职业责任感和幸福感。这次工程实践对本人来说，不但是本年度而且是从业以来感受特别的一次工作经历；言之为文，颇感安慰，愿以自珍之敝帚与业界同人分享与共勉。

注：此文摘引自《城市建筑》2008.12

From the Old Workshop on Campus

Tacit influence. Being a university teacher also a practicing architect, I was often invited to attend the discussion of campus construction. Earlier this year, the school launched its plan to dismantle a group of old workshop buildings on campus at a similar symposium. After I went back to the studio, all the students were overwhelmed on this news. Therefore, the discussion and research on the subject of the settlement of the old workshop was carried out. Also, this symposium lets me initiate a almost one-year design practice and has triggered me the attention and thinking of some "normal" problems at the same time.

I. The Settlement of the Old Workshop

The simple machine and the silence of the workshop are full of thoughts of the past feelings and nostalgia, the great industrial era has ended such feelings, leaving these restless objects making people melancholy...
—— the historian Herbert Harbison

By means of the site investigation and technical consultation of this group workshop buildings, we have the final discussion, that is, to present the *Proposal on the Transformation of the Old Workshop Buildings* which added the new life to the old workshop buildings through transformation. The following three reasons are being illustrated:

1. These remaining buildings are the memorial carrier of the school's development. The school-run workshops were built during 1950s to 1970s and had played a very important role in the combination and development of the industrial-academic-research at that time. From 1990s to now, with the industrial structure transformation and the changes of school's orientation, this group of workshop buildings have become the historical memory gradually, however, they recorded the developing process of the school.

2. Having been reconstructed and designed, the abandoned workshops will become the unique view with special artistic value. These workshops and equipment recorded the former era, including the particular order of the factory, the special space characteristics of the workshops, the industrial and technological beauty models of the equipment and the machines, and so on, and might bring out the distinctive sense of location with little treatment. The old warm red bricks and the new cold grey buildings match with each other make the campus scenery having the characteristic artistic appeal.

3. Having been reconstructed, the abandoned workshops can achieve rebirth and make the suitable application value. "The only luxurious thing here is space, the life here will take all the new look, the space has the interior landscape, the intensions of the architect make them full of vigor just like our city leaving the space open." The existing workshops on campus also have such features, and they are about to become the active place of college with structure reinforcement and facility improvement.

Based on certain persuasiveness, all the explanations of "no taking-down but reuse" in the report are labeled as "ecology". To sum up, on one hand, culture ecology from the historical memory of the campus growth, on the other hand, resource ecology from the point views of the building LCA-design (the Life-cycle in Architecture).

Nowadays, the architectural design is lead by the certain stylish concepts. The architects may be supposed to watch the humble house and their virtue of simple nearby. These groups of buildings on campus are just located on my daily teaching way, but I have neglected them for almost twenty years. Suddenly I drew my attention and found that they have shown numerous qualities being hidden by time : the real shape of space, the real material structure as well as its relevant simplicity, minimalism, and most importantly elegance.

II. The Usage of the Remained Old Workshop

University was originally the market of the idea, where people are able to seek a variety of meaningful ideas.
——*A Pattern Language*

The proposal was mainly discussed in the following aspects:
1.The Most Lacking Space on Campus at That Time
What is the usage of the remained old workshop? And there is a surprising consensus in terms of the answers from the students, that is, what they needed most was the space of the campus culture activities. The layout of the campus was made up of teaching buildings, canteens and dormitories. In addition to rare exchange of ideas in the class of Engineering Colleges, this standard education production line will always repeat itself make it understandable for students to have a strong desire for campus culture activities. While there is still needs the further research if it is risen to the theory altitude.

2. The Purpose and Meaning of Carrying out the Campus Culture Activities

(1) On the Connotation and Basic Object of University Education
In ancient southern Europe, the first object of founding college was that educating students from juveniles with guardians to the grown-ups with abilities of independent thinking and judgment. The focus of education should be people-oriented, and the object of education lies in that enhancing the overall wisdom and morality of the humans instead of the knowledge and technology alone. The basic object of contemporary university education is that: cultivating high-level professionals in general with rational spirit, independent personality and humanities quality. Therefore, such professionals should possess rational spirit, independent personality, humanities quality and professional competence.

(2) The Original Meaning of College and University——The Marketplace of Idea
In colleges and universities, people can pursue eagerly a variety of meaningful ideas. The external environment is supposed to provide the space to encourage rather than to hinder the individual development and the freedom of thought. The surrounding environment must contain the space available for students to understand and distinguish the meaningful thought, at the same time, must offer sufficient opportunities for them to make their own judgment by means of reaching various ideas.

(3) The Significance of EQ (Communication and Expression Ability)
Once giving a lesson to engineering and technical personnel many years ago, Dale Carnegie, the most renowned American adult educator, noticed that the most rewarding are not the people who knew most about the Engineering but the ones who have professional knowledge and are skillful at expressing their intensions and socializing, leading and inspiring the others.

(4) The Campus Culture Activities are the Only Effective Way to Promote Communication and Expression Consciously
The successful campus culture activities can make young students discover themselves, encourage self-education so as to achievement self-concept by means of communication and expression. In terms of technology education, the division of labor of different contemporary disciplines became more specific and the research became more intense which lead to the fact that there is small possibility of cooperation and co-creating between the disciplines, while on the other hand, there is huge potential demand from the necessity of cooperation and co-creating. Therefore, the latest technological, artistic, philosophical and other development of related disciplines must be paid close attention. The premise is that the people engage in various disciplines have the possibility and opportunity to socialize and communicate smoothly and positively.

As a result, this process of discussion ended with the agreement that the old workshop should be renovated into the center of campus culture activities. And its main usages include the following ways: the indoor recreational activities place for faculty and students like artistic gymnastics and ball games, the leisure entertainment place of the teahouse, the book bar, the salon and the cinema, the amateur art school, campus business, the exhibition, the show, and so on and so forth. The center can support the extracurricular life for faculty and students and it is natural to communicate and interact with the suitable space.

With the previous hypothesis and demonstration, we presented a report to the university. The principle was looking forward to the future at seeing the architectural animation of "the center of campus culture in the future" and the marvelous photos of "798" and "LOFT", suddenly he asked me that "would it possible to save money to renovate into the Architectural Hall than newly-build?"

III. From the Old Workshop to Architectural Hall
The architect's responsibility is to provide space for the existing facilities and to provide the possibility for the convenience that is not yet available.
—— Louis Kahn
Throughout this design practice, the master words must be conversed concerning ecology.

1. A Rewarding Graduation Project
Our ideal center of campus culture will be built into an architectural building and all the people associate with architecture will own our own home from then on, which is the desire for the faculty and the students of the Department of Architecture for a long time. But more than excitement, we should confront with the issues facing us at that time reasonably: firstly, our original intention is to build a center of campus culture, what should we do next? Secondly, how to manage and use the luxurious space in the old workshop? Thirdly, how to solve the problem of lacing in funds. This was the moment to select the graduation project, and therefore, a special graduation project emerged.

The teachers of the Department of Architecture illustrated the former problems and relevant thought on architectural building at the beginning of directing the graduation projects. During the process of guiding and tutoring the students' graduation projects, I came up with the key to this question clearer and clearer.

(1) Show Respect to the Sense of Place
The open interior and exterior space design of the Architectural Hall makes it the center of campus culture activities in a spiritual sense. Because the university pays close attention to the construction of the discipline and the Department of Architecture has the opportunity to possess this peculiar gift, we have a deep understanding towards the mission of the Architectural Hall is the center of campus culture activities.

But it just matches the interdisciplinary specialty characteristics of the genus in architecture, that is, the traits here are equal, open and joyful.

(2) Give Priority to the Ecology
We cope with the difficulties elaborately and efficiently from the construction site by means of the rules of the optimum function principle and the initiative principle. This is not only because the funds are not sufficient, also lies in that we have some subjective recognition and active practice. Even later, the functions of the

Architectural Hall are beyond concern, all the efforts are focused on pursing certain and appropriate ecological strategies. This is truly based on the spirit of the humanities of arcology and exemplifies the architects' professional quality of creative problem-solving.

While I was the tutoring graduation projects, I got the previous main guideline to direct my following work. The students who were tutored also experienced the real project process, personally surveying and mapping the old buildings, budgeting and cleaning the reusable components, using and practicing their technical knowledge and training, created the Architectural Hall for their Alma Mater with overwhelming initiative and passion. And I strongly believe that they will have a very different understanding in their work and study later. Being their teacher, I was so inspired from their enthusiasm that I really had an impulse to write and design, also I have deeply felt the pleasure from teaching benefits teachers as well as students up to now.

2. A Special and Ecological Project Practice
Accompanied by the graduation projects, this project practice has entered a stage of full implementation.

As far as the usage of the workshop is concerned, the most valuable thing belongs to the space undoubtedly. Concerning the arrangements of function for existing space and places, the most proper way is to match the adaptability of space and places to the functions that is very different from the normal creation process in which provides suitable space for a certain function and organizes the circulations and partitions for a set of space. Overall, this insertion points lay the ecological foundation for this project practice.

(1) Sense of Place Elicits the Functions
If the location decides the functions, the space efficiency can be played to the maximum extent. When selecting corresponding space for a certain function, without saying, we give special attention to its physical dimension as well as proper atmosphere. For example, the independent workshop in the southeast side must be a wonderful lecture hall due to its sense of order by the beams and sufficient sunshine, some equipment hard to name in the northwest place were displayed according to peculiar order and they would provide a forum to operate, display and illustrate makes it a marvelous place for construction and material display. Combining model manufacture and imagining the students leaning on these old equipment, visiting and operating in different altitude… This environment atmosphere has transcended the common ecological strategy and melted a particular historic memory.

(2) Sense of Place Directs the Functions
One place in the east takes a look of casual scene and breezing coolness, so we add cafeteria to the mandate. There is also a small corner for casting and the firing furnace in it still works, so a new function has directed, that is a "pottery bar" and an optional course has also come into being. The primary strategy of all the ecological strategies is to maximize the space efficiency; meanwhile, the utmost goal of all the ecological strategies is to provide the place and the space with physical and mental pleasure for people.

(3) Create New Place and Space with Memory
During the process of cleaning, the old bricks dismantled of the attached house are used for the materials of the new walls and paving the square; the old material transmission belt is translated into a bridge for teachers and students which can solve the problem of the circulation of the independent southwest workshop and the main space. At the same time, the original nature of the factory is brought into the new space; and all composite steel columns of the transmission belt supply the total steel material for the new floors; all the steel window materials are converted to the rails; the giant reinforced concrete crane beam turns into the stacked chairs in the Lecture Building and the stools inside of the Architectural Hall; the bare reinforced concrete bracket naturally becomes the best supporting corbel of the upper structure beam-column, and it contributes to a very unique space in the building, that is, the space on the top, which is different from the modern concept of storey height. The remained machines are put outside, however, suddenly, they turn into and more like the modern art installations of creating atmosphere, recalling memories…

In this case, we take other ecological strategies, such as structure reinforcement and functional layer. The chimney is turned into ventilation equipment. Skylight can balance wind and light. What we got is ecology is neither established procedure nor slogan. It is active behavior with enthusiasm. I hope everyone has the experience and I'm so lucky to do this with my classmates.

I had different feelings to building in the discussion with teachers and classmates. I was calm when I stood in the old workshop which had limited money and high expectation. I was so glad when I got a new idea in the construction site. I felt so proud to be an architect. This case is a special experience not only for this year but for my career. I hope we can share these together.

Note: It quoted from *Urbanism and Architecture* in December, 2008.

一次务实的旧厂房改造与再利用实践

【摘要】通过对内蒙古工业大学中一处废旧厂房的改造实践，试图总结一针对此类厂房改造与再利用的务实模式，包括以下四个方面的策略：1）空间的有效利用：评价厂房的空间特征，适配最佳的使用功能，以最少的用功创造最大化的空间使用模式；2）对症式的结构加固：采用一种抗震性能评价与功能再造相结合并加强薄弱部位的加固模式；3）被动式的生态通风系统：利用厂房空间高而开放的特点，结合天窗、烟囱、地道等因素组织自然的室内气流，在达到通风、降温、加湿等效果的前提下，保证室内空气的质量和环境舒适度；4）废旧材料的重新利用：利用厂房的废旧材料和构件，转换功能易位再用等。

【关键词】旧厂房；改造；生态；策略；务实模式

一、背景

在倡导低碳经济和推行绿色建筑的背景下，建筑行业内普遍将视点锁定在建筑资源再利用的研究和实践上。国内可再利用的建筑资源基本有三类：既存住宅、工业遗产和历史建筑遗存。其中的工业遗产，特别是20世纪50年代工业产业的快速发展，在城市重要区域中留下了相当数量的废旧厂房。这类厂房在三类建筑资源中具有较强的特征：坚固的结构构件、高大开敞的建筑空间、寿命周期较长的建筑材料等，这些特征为它们今后的改造与再利用提供了可能。

针对此类厂房的改造国内外均有范例。国外方面成功的案例有英国的泰特博物馆和法国的维尔茨堡美术馆等，它们改造前均为废旧厂房，改造的核心目的是保留遗产，改造的成本普遍较大甚至超过重建，不适合中国的国情；国内近年也出现了类似的改造活动，从自发、零散到作秀、示范再到有理念、有策略、用较尖端的生态技术（双层表皮、自动遮阳）等，集大成者可推上海世博的旧厂房改造项目，但整体考察其改造模式认为仍不够务实和全面：或保留甚少，或造价昂贵，或为临时性的用途等，其示范引导意义大于实际推广价值。

2008年初，我们申请改造内蒙古工业大学校园中的一处废旧厂房，赋予它新的功能。该厂房曾经是学校教学实践基地的铸工车间，1968年建设，1971年投产，1995年全面废弃。厂房整体上保留完好，其空间的通透、结构的自然、细部的率真等都让我们为之着迷。激情之余，伴着上述思考，改造设计试图总结出一些务实的策略和模式，并把这种设计策略和模式的产生与改造施工的全过程有机结合起来。

二、策略

设计伊始，经过慎重比较，定位为建筑学的教学场所——建筑馆，改造的策略表现在以下四个方面：

1.空间优先的设计策略

寻找最佳功能即为第一个设计策略。区别于根据使用功能设置空间的一般模式，采取空间优先的设计策略。即，对厂房的空间特征进行适应性评价，在可能的范围内赋予最适宜的使用功能，以最少的用功创造最大化的空间使用模式。这是一种着眼于生态集约的设计操作。在内蒙古工业大学的旧厂房改造项目中，首先，针对其空间开放的特点，定位为重交流、重体验、重实践的建筑学教学场所——建筑馆，否定了作为其他学科的实验场馆以及因需分隔不符合绿色初衷的校园文化中心（音、体、美的功能彼此干扰），并强烈地意识到厂房通透开敞的大空间、自然裸露的结构构件、不加掩饰的构造细部本身就是一个"天然"的建筑馆；其次，根据各处空间的大小、尺度、明暗及氛围等布置阅览、评图、展览、报告、设计以及模型制作、艺术沙龙等具体的功能房间。这种空间优先的设计策略不仅表现在功能对于空间的"适应"方面，还表现在空间对于功能的"诱发"作用。

基于"适应"布置的功能房间有：

报告厅——东南角的独立车间容量适宜、视线和声音俱佳，顶部的天车正好作为灯光和投影机的桥架；

图书阅览室——底层面向院落的一处空间安静而阳光充足，置身其中还可以观赏到院中的大树；

美术教室——顶部的空间南向无窗，来自于天窗的光正是其最佳的稳定光源（美术天光教室）；

计算机教室——形体转折处的房间较暗，正是屏幕阅读的理性光线，而隔壁保留的"冲天炉"为其解决了通风问题；

入口大厅——两个主车间的连接处，同时处于院子的中央，便于人流的集散和内部动线的组织；

还有，北部的小房间成为厕所，梯下的房间成为图文数据室，靠近路边的工具室成为管理门房……

来自"诱发"的新功能房间有：
模型制作室——西端的独立车间是整个生产线的起始端，保留的机器充满了制作感，独立的出入口为原料、垃圾提供了方便，是一个理想的模型制作室；

沙龙——东侧的车间，废旧的机器、纵横的管道充满了酷意，是一个天然的艺术沙龙；

还有，院中拆除焙烧房墙体裸露出的炉架演绎成一处休息空间；东侧的料场由于周围路面的逐年抬高自然设计成一个下沉的室外展场……

当然，整个空间利用的过程都是基于开放的原则，这不仅来自建筑学教学的特点，更源于省功省力的生态前提。

2.综合多效的结构加固策略

旧厂房层高较大，跨度也大，结构承载力高于一般民用建筑。具有坚固、耐久的主体结构和结构受力简单明了的特点。但是经历了较长时间的使用，它的牢固性和使用寿命都有所降低。

针对厂房结构特性，结合功能的再利用方式，采用抗震概念设计与加强薄弱部位相结合的加固模式。具体做法是：在考察与评价结构的抗震性能以及与现有规范差距的前提下，首先，寻求一种综合多效的解决方案。如，加层是获得面积最有效的方式，把这种加层实施在结构侧向刚度和稳定性最薄弱的部位，使新加结构与原有结构形成整体，完善了支撑系统，提高结构的刚度和整体延性。再如，合理封堵原厂房的大窗户是节能的有效措施，把这种封堵与加强墙体刚度结合起来，能够有效地提高墙体的整体抗震性能等；其次，在上述前提下，对改造后仍然存在的薄弱节点处实施局部加固。如，扩大局部柱截面、增大屋盖梁的支承长度等。

3.绿色生态的通风策略

通风有主动和被动两种：主动，耗电耗能；被动，则绿色生态。利用厂房空间高而开放的特点，组织被动式的室内气流是本设计的主要策略。此类工业厂房均包含天窗、烟囱、地道等组成部分，它们构成了潜在的自然通风路径，巧妙地利用风压和热压的共同作用加以组织，不但实现通风，还可降温、加湿，保证了室内空气的质量和环境舒适度。在内蒙古工业大学的旧厂房改造项目中，通过计算天窗的开启面积、测量地道的宽度、确定进风口的合理位置、组织与室内地道的连通方式以及顶部天窗设置电动开启扇等系列措施，使室外的新鲜空气经地下通道冷却后进入室内，再经可电控的天窗向上提升，组织了开放空间的自然通风的同时，通道入口处设置的水池又带入了湿气。保留的烟囱则组织了封闭的报告厅、物理实验室等房间的通风。

4.废旧材料的再利用策略

在整个厂房的改造过程中，精心利用厂房的废旧材料和构件，转换功用易位再用，这不仅是一种节能节材的方法，同时也是一种保留人文记忆的有效策略。如，在内蒙古工业大学的旧厂房改造项目中，将原有生产线的钢柱和钢梁，切割后用作加层的主要构件；将大量的天车导轨变成钢梁、钢柱、钢梯的主材；裸露的钢筋混凝土牛腿柱经过计算成为上层结构梁的最佳依托；旧厂房大面积的开窗留下了数量巨大的钢窗废料，将其切割、焊接，变成楼梯和共享空间的栏杆；废弃冶炼炉拆出的耐火砖，完整的用于室外庭院的铺地，破碎的做了地面基层；旧机器适当处理转换为室内外的艺术装置等。在大量实施上述策略的同时，还将拆出来的零星废旧钢板拼成自由图案用于门窗的合理封堵等。

上述四个方面的改造策略在实施中并非孤立地进行而是相互叠加的，还需要说明的是：它们的实施都有赖于艺术的加工和处理，非此，则难以创造人性化的环境质量和具有个性的场所氛围，因不是本文的重点，此不赘述。

三、性能

通过上述策略，本项设计追求如下三方面的综合性能：

1.质量—经济的平衡性

质量—经济的平衡性是获得质量所耗费资源价值的综合度量，在质量相同的情况下，耗费资源价值量越小，其经济性就越好，反之则差。国内外不同废旧厂房的再利用项目，由于改造理念不同，耗费资金量差异很大。如英国的泰特美术馆、上海世博的某些工厂改造项目等耗费资金较大，质量虽高，但显然不是本项目改造的目标；再如北京的"798"项目造价虽低，但其综合性能不高，也非本项目所追求的目标。本设计在内蒙古工业大学废旧厂房的改造过程中，追求付出资金与获得价值的综合性能比，以求得到最佳的质量—经济性。

2. 系统改造的整体性

本项目改造注重综合的质量，注重各设计元素之间的关联并形成系统改造基础上的整体性。具体表现在：内部与环境的整体统一、功能与空间的整体统一、氛围与动线的整体统一、加层与加固的整体统一、保护与艺术效果的整体统一以及改造与人文记忆的整体统一，也包括造价与质量性能的整体统一等。

3. 生态策略的全面性

本项目改造较全面地使用了多种生态策略，并力争创造性地和旧厂房的改造内容结合起来，尤其注重使用低技前提下的适宜策略。具体分为单一策略和多效策略两个方面，总结如下：

单一策略有：

1）利用原有天窗、烟囱、地道等构筑物组织室内气流的路径，达到通风降温的目的；
2）利用原厂房天窗进行天然采光，节约了照明能耗；
3）利用废旧材料，节约了材料和运输的费用；
4）采用低温地面辐射采暖加明管的方式，节约了采暖能耗；
5）大量使用钢、玻璃等循环建材，提高了材料的长效利用率；
6）保留树木，增强生态效应。

多效策略有：

1）加层和加固配合进行，节约了加固的费用；
2）封堵窗洞和保温措施协同进行，节约了保温的费用；
3）加固、封堵、保温同时作用，又节约了各自的费用；
4）功能划分与空间特征相配合，省去了完善功能所需的用工；
5）廉价材料与场所氛围相结合，节省了材料的造价；
6）增加室内湿度的水池循环用于浇树，节约了水资源；
7）防寒门斗、阳光房与交流场所有机结合，既保暖节能又增加气氛。

需要说明的是，适宜技术的叠合应用本身即是一种统合多效的策略。

四、效应

通过这些节能、节材的改造策略及其产生的综合性能使得预期的两种效应十分明显：一是强化了原有场所的工业气氛，保存了一种特定的历史人文记忆；二是利用了废材、减少了垃圾、节约了造价。对于后者，在内蒙古工业大学的旧厂房改造中，废旧材料占了整体用材90%的比例，加上空间、结构、通风等其他方面的策略，建筑造价每平方米仅用了600元左右，节约造价每平方米1500元左右。本项目原有建筑面积2746平方米，改造后获得面积5960平方米，节约造价近900万元。

经旧厂房改造建成的建筑馆于2009年5月投入使用。现已成为校园内学习环境优雅、学术氛围浓厚、交流自由开放、人文记忆深刻的场所，成为内蒙古工业大学对外交流宣传的窗口，受到业界的普遍好评。目前，建筑馆报告厅已成为学校主要的学术交流场所，每年举行近百场报告会和其他活动；展览空间定期举办建筑、规划、艺术类专业师生作品展以及区内外高校、设计院的作品展。

值得一提的是，在建筑馆（旧厂房改造）项目获得成功的基础上，内蒙古工业大学另外两座旧厂房的改造和利用工作已开始实施，并将于年底竣工使用。

注：此文摘引自《建筑技艺》2011.05

A Practice of Practical Rebuilding and Recycling of an Old Factory Building

Abstract: Through the reconstruction of the waste plant in Inner Mongolia University of Technology, we try to explore a more practical model for reuse of the buildings. Mainly there are four strategies: 1. Effective use of space, which means assessing the space features of the plant and endowing the best functions in order to be space-saving and low-costing. 2. Structural consolidation, which means combining the evaluation of the anti-seismic performance and consolidation of the weak areas. 3. Negative ventilation system, which means combining dormers, chimneys and tunnels to lead indoor air in order to reach the goal of ventilating, cooling and humidification and create a comfortable environment and ensure the quality of the air. 4. Reuse of the waste materials, which means making use of the waste materials and components in order to change their functions and reuse them in different places.

Key Words: Waste Plant, Reconstruction, Ecology, Strategies, Practical Models

I. Background

Against the background of advocating low-carbon economy and promoting green buildings, general perspectives focus on research and practice of recycling of the building materials. Domestically there are three categories of the recycling building materials: existing residential, industrial heritage and historic remains. The industrial heritage includes, especially after the rapid development of industry in the 1950s, a considerable amount of abandoned waste plant in the center of the cities. Those waste plants have the peculiar features as consolidated structures, grand spacious scales and endurable building materials, which make it possible to recycle and reuse the deserted plant.

There are many examples of the recycling of the waste plant both at home and abroad. Internationally the most typical examples are the Tate Museum in Britain and the Wuerzburg Art Museum in France. As the former waste plant, the recycling of those museums was costly, which was not suitable for China. In recent years, China also has the similar attempt, ranging from individual exploration, different samples to the strategic use of high-tech and eco-tech. The rebuilding of the waste plant of Shanghai World Expo is the most well-known and successful one. However, the project is still considered not practical and overall, for it maintains some costly and temporary functions, which means the project leans much more on its promotional value than on its actual value.

In early 2008, we applied for reconstructing the waste plant in Inner Mongolia University of Technology to give it a new life. As the former founder workshop for internship, the plant was built from 1968 to 1971 and was abandoned in 1995. We are obsessed with its complete framework, spacious rooms and fascinated details. And with those obsessions we try to combine the practical strategies with the whole process of the reconstruction.

II. Strategies

At the very beginning, the waste plant is planned to be reconstructed as the teaching building of architecture. The reconstruction strategies are as follows:

1. Priority of Space

The foremost strategy must be searching for the best function. Unlike the approach of prioritizing the space of the building, finding the most suitable function comes first, which serves as a typically Eco-intensive design. As for the project, in consideration of the grand space and the exposed building blocks of the plant, we emphasize the communicational, experiential and practical features of the reconstructed building, neglecting its function as a lab spot and as a gymnasium on campus. Additionally, we also design the specific functional rooms for displays, presentation, design, model making and art salon in accordance with the size, lightening and atmosphere of the space. In all, the strategy lies on not only its adaptive value but also its stimulating value.

The rooms of adaptive functions are:

Auditorium/ the Lecture Hall-the independent workshop in the southeast corner, which has a good capacity, lights and sounding. The crane on the top can be used as the bridge of lightening and projector.

Reading Room- the corner facing the garden on the first floor, which is tranquil and sunny. It has a good view of the big tree in the garden.

Painting Room- the top of the building, which has no windows. The lights come from the dormer.

Computer Room-the dim turn of the building, which is suitable for screen reading and the neighborhood cupola is good for ventilation.

Entrance Building-the junction of the two main workshops, which is located at the center and good for distribution of the flow of people and organization of the internal moving lines.

Also, the small room on the north as a washroom, the room under the stairs as reference room and tool room by the road as concierge…

The rooms of stimulating functions are:
Room of model making-the independent workshop on the west is the beginning point of the assembly line. The remaining machines remind people of hand making process and the independent entrance is convenient for raw materials and trashes.

Salon-the workshop on the east, which is a cool place full of deserted machines and scattered pipes.

Also, the exposed hobs in the garden as a resting place, the stockyard on the east as a sunk outdoor display room…

The use of the room is based on the principle of openness, which complies with the teaching traits of architecture and the ecological labor saving principle.

2. Consolidation of Structure

The carrying capacity of the waste plant is much greater than the ordinary residential buildings. The main structure is stabilized and endurable and its force structure is simple and concise. But after the long-time use, its stability and service life are reduced.

In accordance with the structural features and the reuse of the functions, anti-seismic design and consolidation of the weak areas are fully used. To be specific, after the careful assessment of its seismic performance, first of all, a comprehensive and multifunctional approach is adopted. To get more space, the most effective is to create more floors. The consolidation is made on the structure where the structural lateral stiffness and its stability is the weakest, which combines the former structure and the new structure perfectly and enforces the supporting system. Furthermore, sealing the big windows is also an effective way for enhancing its seismic performance. Additionally, after the consolidation of the main structure, the repeated reinforcement of the afterward structure is also needed, such as, expanding the fragmental column sections and prolonging the roof beams.

3. Ecological Ventilation

Ventilation consists of two kinds: positive ways, which means consumption of electricity and energy and negative ways, which is environmental friendly. And negative ventilation is adopted because of the height and the openness of the plant. The dormers, chimneys and underground tunnels serve as the natural ventilating paths. The combination of wing pressure and thermo compression not only has the function of cooling and humidification but also ensures indoor air quality and environmental comfort. In the project through calculating the open area and breadth of the tunnels, assuring the location of air inlet, organizing the connection of indoor tunnels and installing the electric fan on the ceiling, the outdoor fresh air will be cooled and led to the building and will be elevated by the electrically controlled dormer to make it circulate. At the same time, water vapor will be created after the air passing by the pool of the air inlet. The reserved chimneys keep the ventilation of the enclosed auditorium and physics labs.

4. Recycling of the Waste Materials

In the process of the reconstruction recycling the waste materials and artifacts is not only a way of saving materials and energy but also an effective way of memorizing the histories. For example, the original steel columns and steel beams are cut and used as main structures of adding floors. Large amounts of cranes are transformed in to steel beams, steel columns and steel stairs. Exposed reinforced concrete cow-leg pillar is used as structural columns of the upper layer. Many waste steel window frames are cut and welded to be used as handrails. Refractory bricks from the smelting furnace are all reused as the floor of the garden. The old machines are slightly changed into the indoor and outdoor art equipments. Last but not the least, the waste steel plates are made into different patterns to block the dormer.

The strategies are overlapped with one another when adopted. What needs to highlighted is that the reuse of materials depends on artistically processing; otherwise, it will be difficult to create a humanistic environment and fascinating surroundings. This will not be repeated here.

III. Performance

The project will reach the following three performances:

1. Balance between Performance and Cost

Balance between quality and economy is the measurement for consumption of materials. The less the consumption of materials is, the more rewards the project makes, and vice versa. The reconstruction cases both at home and abroad spend money differently due to different strategies. The Tate Gallery in Britain and the Shanghai World Expo are all costly. And on the other hand, the Beijing 798 project is low-cost but low-performed. The project of waste plant in Inner Mongolia University of Technology is designed for balancing performance and cost.

2. Global Views of the Reconstruction

The project lays emphasis on integrality of the performance, connecting different designs, to be specific, the integrality of indoors and outdoors, functions and space, atmosphere and moving lines, floor and consolidation, reuse and artistic effects, restoration and histories and costs and qualities.

3. Global View of Ecology

The project makes use of many ecological strategies, which combines the creativity with the reconstruction. The strategies can be divided into single ones and multifunctional ones.

Single strategies are:

1) Using the former dormers, chimneys and tunnels to lead ventilation in order to cool the air;

2) Using the former dormer to get daylights in order to save electricity;

3) Using waste materials in order to save building materials and cost of transportation;

4) Using low-temperature radiant floor heating plus heating pipes in order to save the consumption of heating;

5) Using the recycling materials of steel and glass in order to

increase the long-time use of materials;
6) Keeping the big trees in order to reinforce the ecological effects.

The multifunctional strategies are:
1) Combining adding stories with consolidation in order to save the cost of reinforcement of structure;
2) Combining blocking the windows with insulation measures in order to save the cost of heating;
3) Combining consolidation, blockade and heating in order to save the overall cost;
4) Combining functions with the space in order to save the time of perfecting the functions;
5) Combining low-cost materials with creation of the atmosphere in order to save the cost of materials;
6) The water from the pool used for both increasing the indoor humidity and watering trees in order to save water;
7) Combining warm cross-bars, light rooms and communication rooms in order to keep warm and save cost.
What needs to be highlighted is that the overlapping use of proper techniques is also a multifunctional strategy.

IV. Effects

Through the adoption of energy-saving and material-saving strategies and the prospective performance, the two effects of the project are obvious. The first one is the reinforcement of the industrial atmosphere of the original spot and memory of the specific historic and cultural heritage. The second one is the recycling of waste materials and reduction of cost. As for the second one, in the project waste materials account for 90% of the overall materials. With the strategies of space, structure and ventilation the construction cost of per square meter is 600 RMB or so, the saving cost of per square meter is 1500 RMB. The overall construction area of the project is 2746 square meter and the reconstruction area reaches 5960 square meter, saving cost of 9 million RMB. The architecture Building has been come into use in May 2009. It is now a place for promotion of the university due to its elegant learning atmosphere, free academic atmosphere, open communication and memories of the histories. At present, the auditorium is the center for academic interactions, holding hundreds of lectures and seminars and other activities each year. And the display Building holds exhibitions of teachers and students from different universities and designing institutes on the regular basis.

What needs to be mentioned, after the completion of the project, the reconstruction and reuse of another two waste plants of Inner Mongolia University of Technology has been launched and will be completed at the end of the year.

Note: It quoted from *Architecture Technique* in May, 2011.

三、关于建筑馆的评品
III. The Evaluation of the Architectural Hall

赵 辰 建筑学的力量
——从内蒙古工业大学建筑馆看工业遗产保护之建筑学主体意义

摘要： 工业遗产保护与再利用已成为当今社会发展的重要内容，建筑学专业必然要为之做出相应的贡献，但目前业界还存在不少误解。内蒙古工业大学建筑馆作为一个成功改造自原废弃机械工厂的成功案例，透彻地说明了在改造过程中，不能简单地注重工业遗产的保护，而应强调工业建筑的再利用，将工业建筑重新融入城市空间环境，从而充分地显示建筑学的主体意义。

关键词： 工业遗产；误区；工业遗产保护；再利用

内蒙古工业大学建筑馆，建筑学主体意义内蒙古工业大学建筑馆（下文简称"建筑馆"），2009年改造自校内一处业已废弃多年的机械厂房。这一保留了原厂房内、外环境基本空间形象特征的工业建筑，现已成为内蒙古工业大学建筑学院教学、科研的场所。看到这个破败凋零的荒废之地经"华丽转身"而成为建筑学的学术殿堂，观者无不为之喝彩。笔者也深有同感，但同时认为：以这一成功的工业建筑保护改造案例，结合当今工业建筑遗产保护与再利用的社会意义，其中仍有一些深刻的建筑学术意义值得探讨。

工业建筑文化遗产保护，在当下已是一个很热门的社会事业和公众话题，社会各界都投入了相当的关注度。一些著名的保护性改造利用"成功案例"，更是产生了极大的社会反响。笔者近些年来也常关注此课题，在国内外也考察过一些重要的案例，并参与过相关的实践性项目研究工作。通过这些研究，笔者充分认识到：在社会各界的关注与支持下，这项事业在积极发展的同时，也存在不少误解。某些认识论与方法论上的问题，必将对其发展产生不利影响，对建筑师来讲尤其严重。笔者将之定义为："工业遗产保护的建筑学误区"，并曾在一些会议上发表过此见解。

而张鹏举老师主持下进行的"建筑馆"项目的成功，则与笔者针对这些"误区"所开展的讨论中的观点完全吻合。反映了作者坚持建筑学的主体，也体现了建筑学的意义。为此，笔者希望通过此案例的实际情况，来进一步阐述工业遗产保护中的建筑学意义。

一、工业遗产保护作为建筑学的新内容
首先，作为建筑学的一项新内容，对工业遗产保护和利用的基本概念进行讨论。

工业遗产保护与利用，在一些发达国家已有数年的发展历史，尤其是英、德等工业革命源头之欧洲国家。在我国，也有10年以上的发展过程。由于人类文明发展的阶段性原因，社会正走向后工业、信息化、城市化时代。这种社会的转型，使得原来曾为社会生产大量效益的工业厂区，出现了停产、转产、迁移等行为。于是，大量原来的工业建筑物及所占的用地区域改变用途，甚至闲置、荒废。如何将这些建筑物和用地区域重新利用，就成了建筑学的新课题。由于中国的社会高速发展，使这种现象来得尤其迅猛，而建筑学的应对则显得仓促。

作为一个社会历史发展的新理念，目前在国际上已经有比较严格和清晰的概念界定。主要是由"国际工业遗产保护委员会"在"国际古迹遗产理事会"的框架下，于2003年制定的"下塔吉尔宪章"（The Nizhny Tagil Charter）。其中相关的文字定义十分明确，强调了"凡为工业活动所造建筑与结构、此类建筑与结构中所含工艺和工具以及这类建筑与结构所处城镇与景观，以及其所有其他物质和非物质表现，均具备至关重要的意义"。我国的国家文物局、住房建设保障局、文化部也曾发出了关于工业遗产保护方面的如"无锡建议"等指导性文件。从中国的情况来看，相对于发达国家我们的起步较晚，很明显，任务也更为艰巨。由于这些新理念的"宪章"和"文件"所提出的要求保护的工业建筑遗产，涉及的领域相当广。尤其是从"下塔吉尔宪章"等提出的概念来看，这种工业遗产与其他历史文化遗产在定义上的差别不大，要求保护的力度自然也不小。我们很容易顺理成章地以为：工业文化遗产保护应等同于其他历史文化遗产保护。于是，作为建筑师来讲，除以前需要参与保护文化类的历史建筑以外，又要增加对工业遗产建筑的保护工作了。

事实上，这些"宪章"或"建议"大都是讨论有关保护这些工业遗产意义的，而建筑师更多的工作是关于如何去做的问题。从这个层面来看，这些"宪章"和"建议"以及相关的文化界人士们的讨论，能帮助我们的并不多。规划与设计的问题又是严重的，工业遗产的内容与其他的文化遗产，在空间特征上有很大不同。"下塔吉尔宪章"提到："工业文化遗迹，包括建筑和机械、厂房、生产作坊、工厂矿场以及加工提炼遗址、仓库货栈、生产、转换和使用的场所，交通运输及其基础设施以及用于住所、宗教崇拜或教育等与工业相关的社会活动场所。"这些建筑物或是构筑物往往都是尺度巨大、体积粗犷，且包括了相当多的机械与器械。这样的空间场所，难以与常人的生活、工作等活动相关联。简单地说，工业遗产，在建筑特点上具有一种机器化而非人性的尺度和空间。尽管工业建筑的空间与场所作为工业文化的遗产，应该得到一定意义上的尊重和保护。但是，要将这些人类工业文明造就的一些巨大的"怪物"，像保护其他文化类历史建筑遗产那样来对待，是不大现实的。

"建筑馆"，无疑是属于这种作为建筑学新内容的工业遗产保护的案例。而此建筑的成功之处，正在于该项目的主要建筑师张鹏举以及项目决策者、执行者们，并未将之简单的作为文化遗产的保护工作来对待。然而当今业界存在着相当多简单地以延续文化遗产保护的思想方法，来对待工业遗产的问题。笔者以为，这恰恰是一种误解。

二、工业遗产保护成为建筑学的误区

面对工业遗产这种新的社会现象，我们需要引入一些新的、关于建筑审美的价值观。这意味着，大可不必将原本粗犷的工业建筑的空间场所，加以过多的修缮或是装饰。

要接受"机器化"的粗野美感，让局部的人性化尺度的活动在整体的工业化尺度空间中实现，已然成为我们作为建筑师的一项重要任务。不过笔者也怀疑这是否是我们建筑师的主要工作，因为我们能够看到的是，某些摄影师、设计师、记者或是媒体制作者，完全有能力通过一些"秀场"来做到这一点，而且往往效果比建筑师的工作要更好。

作为建筑师，面对新的社会需求，必然要有新的理念和方法。若是沿用对待文物类建筑所习惯的观念和方法简单地强调在建筑设计中的严格保护，则必然带来不可避免的误解。从建筑设计的工作性质来看，误区主要反映在两个层面。

1.误区之一，保护还是利用

我们似乎过多地谈论了工业遗产的保护问题，而忘却了与之直接相联的利用问题。事实上这两者在建筑学的意义上是密切联系、不可分割的，不与利用相联系的工业遗产保护，大概并不存在，即便存在也与建筑师的工作关系不大。

对于需要承担具体工程项目之规划与设计工作的建筑师来讲，比在理论上讨论保护工业遗产的重要性还要重要的问题，是如何去合理地利用这些"遗产"。即：要研究、规划、设计这些工业建筑空间的再利用。大部分情形下，建筑师需要在已废弃的工业建筑与环境中，重新置入新的程序与功能，要合理安排新的活动内容。这意味着其设计受到了极大的限制，已经不可能是简单的"形式追随功能"之"功能主义"的造型。

"建筑馆"作为一个实际的工程项目，正是立足于如何利用原内蒙古工业大学校内机械工厂之现实问题。该厂房废弃多年后，学校原本希望通过拆除而使用此用地。此时，张鹏举等正是出于其建筑师的敏感，提出了可以不拆除而通过改造来利用的合理化建议。这个建议，可以让学校在相对较少的投资条件下，解决建筑学院的教学、科研空间场所。而学校领导也在采纳了其建议后，顺水推舟地将之变成建筑学院的场所改造任务。这里的背景条件是，内蒙古工业大学建筑学院原本并无自己的独立场所，与建筑工程学院合用一幢楼，空间十分局促，很难达到全国建筑学专业教学评估所要求达到的空间场所标准。

很显然，利用废旧机械工厂的改造来得到有效的建筑学院空间场所，成为这项工程的基本目的。利用，是作为出发点的。

当然，合适的利用必然意味着一定意义上成功的保护，建筑学的意义从来就在于整合、协调各种因素，并不是某一单项因素的纯粹意义。"建筑馆"的成功恰恰是结合利用，成为建筑学院保护工业遗产的案例。也是最充分的、最有力的利用为先、兼顾保护的实践证明。

从职业性质来分析：面对工业建筑遗产这样的社会问题，在工作的性质上应该有社会分工的不同；对文物工作者来讲自然是保护问题，这也是社会文化界的基本价值导向；而对作为承担规划设计任务的建筑师来说，则基本是以利用为核心的问题。当然，我们希望是有保护的利用。但是，利用应该是更优先的。当保护和利用有矛盾，二者不可同时兼备时，还是必须拆除一部分。将工业遗产如远古遗存的文物建筑一样全面保护是不可能的，也是不现实的。尤其对于那些实在已经不能再利用的工业建筑物与构筑物，若是依然过分地强调保护，只能导致社会的不合理发展。对这些问题的研究与思考，正是以建筑的设计与规划工作为特征的。从社会的职业与专业分工角度来反思，如果缺少了建筑师去集中思考再利用问题，必然会导致整个工业遗产事业的不利。

不少建筑师在面对工业遗产任务时，忘却了作为建筑师的职业责任和建筑学的专业素质，仅仅是跟着文物学者、文化学者一味地高谈保护。似乎以此占有了时代的潮流、社会的风尚，殊不知，建筑学的价值将随之降低或丧失。更为不利的是，我们建筑学术界也如此这般。只强调保护并以此来教育年轻一代的建筑师们，当我们将建筑师们都鼓动成了如同社会文化学者一样，只关注保护工业遗产而不懂如何利用工业建筑空间场所时，从社会角度来看，当建筑师只会唱保护的高调而不具备利用的能力的话，这种"保护"事业必将失败。

令人欣慰的是，"建筑馆"案例，就是以利用为核心的。笔者更为高兴的是，"建筑馆"作为建筑学院，让内蒙古工业大学的莘莘学子能耳濡目染地体会这一工业厂房的合理利用，使他们能够正确理解工业遗产在今天的意义和建筑学的意义。其实，这样的成功例子国内外有很多，其核心都是利用。如德国鲁尔工业区之"关税联盟"（ZollVerrein）、英国利物浦港口"阿尔伯特仓储区"（Albert Dock）等著名的项目，都是以利用为主的成功改造案例。其文化遗产的意义都是在改造成功以后，才真正彰显出来的。而中国一些所谓著名的案例，更是在学者、官方都未意识到工业遗产保护意义之前，一些有敏感度的艺术家、设计师们，主动地对行将就木的老厂房与仓库合理利用，改造为工作室、事务所等。如著名的北京"798"和上海的"苏州河"沿岸，并不是什么高深的保护理论之成果，而是实际的利用行为之产品。

2.误区之二，工业复兴还是城市复兴

我们似乎也过多地谈论了工业的问题，而忘却了城市或环境的问题，这正是笔者以为的第二个误区。

大部分的工业建筑或遗产都与更大范围的城市环境相关联，当我们对一个工业区域进行保护与利用的规划时，必然碰到所谓"保护"与"复兴"的概念。然而，对于一个工业建筑所占有的区域来讲，保护和复兴到底是针对工业，还是针对城市或者环境呢？很多时候，这个问题被混淆了。笔者曾参加过以此为主题的国际工作营的评图，其主持者提出的命题是"工业复兴"（Industry Rehabilitation），这恰恰完全误导了参与者的设计研究方向。从其中某些作品中的问题，更是明显地反映出了这种理解上的错误，一些设计在极其关照工业建筑象征性要素的同时，完全不考虑整体的厂区如何与城市各个部分的联系，尤其是交通与流线。特别是中国国有企业体制的僵化和土地管理制度的特殊性，导致不少企业在工业厂区改造为"创意产业园"时，完全不考虑周边城市环境而固守自己的圈地范围。从城市发展和环境问题角度来看，这实际上十分有害。这种情况在近些年的许多项目中都有所体现，甚至于所谓的著名项目。如南京的原晨光机器厂及金陵机器制造局厂址所进行的"1865创意产业园"，自2003年以来，整体的开发已有相当力度；但是，在与城市关系方面考虑相当不足，也未受到相关部门的关注。一旦深入发展，与城市相关联的交通、景观、业态等方面问题必然不断暴露，且难以解决。这正是过多地只是谈论工业遗产的保护，忘却了也是更为重要的城市复兴问题所导致的，也是建筑学术界长期以来的一种避重就轻的态度所造成的恶果。即便是"人人称道"的北京"798"，在笔者看来，在城市复兴意义上完全不能算是成功的。同样存在着交通流线不便、城市空间场所标示性差，并隐含着相当多的城市公共性与开放性方面的问题。

我们应该认识到，工业本身的保护必要性应有限度，且没有"被复兴"的必要。从社会的角度来看，工业消亡或迁移了，这是社会发展的规律。真正需要复兴和保护的对象，应该是城市，或者说是环境。在城市中，原有工业建筑区域复兴后如何与城市环境相适应？工业区原本就是城市的一部分，往往还是其重要的经济区域。社会的发展过程使工业和产业发生衰落或转移，从城市的角度来看，意味着这个原工业区域重新被城市化。

从"建筑馆"来看，这个层面的问题主要反映为，如何将原来与环境极不协调的生硬粗野之厂房，重新融入校园中。我们可喜地看到，"建筑馆"在与校园相关联的接口，做出了十分积极的外部空间设计；在"建筑馆"的院落内外，有大量利用原工厂设施或构件改造而成的院门、天桥、栏杆等；并向院外增设了室外会场、景墙、座椅等外部空间设施，使建筑学院的部分空间向校园开放、延伸。这些措施使得这个项目的建筑空间不是停留在其本身和内部，而是与校园环境整体呈现一种合作、相融的态势。随着下一期扩建工程的建设，可以预见将进一步与校园环境达成协调而复兴的发展景象。

"建筑馆"所采取的对周边环境主动积极的空间融合态度，正是当下大量的工业遗产保护改造项目中所缺少的与城市和环境合理互动的关系思考。工业遗产所在地的城市环境，只有在改造项目的规划设计中充分考虑相互间在空间上渗透、互动、交流、融合，尤其是向原工业区域充分引入城市公共空间，才能做到城市区域的真正复兴。

三、以建筑学的主体性重新审视工业建筑遗产问题

建筑学学科与职业的综合性特征，反映在不断面对新的社会问题时，其研究领域会得到拓展。这自然意味着建筑师们要了解和学习一些新的东西，包括新的理论、新的设计、新的样式等。而这些新的内容往往都有其他学科、专业、领域的特征，似乎我们必然要涉入其他领域，或者是要受其他学科影响。但是，如果我们只顾所拓展出去的那些东西的意义，而忘了自己学科与专业的本质，那建筑学就容易失去自己的主体性。建筑学是必须不断面对这些问题的，所谓建筑学的定义，应该是一个永恒的问题。从逻辑意义上来定义，建筑学作为一种概念的内涵和外延有以下特征：其外延有比较大的不确定性，这将导致建筑学不断会有新的内容参与；但是其自身的内涵具有相对的稳定性，也即不论有多少新东西要面对，建筑学都要创造性地去解决社会问题。建筑学的相对稳定之内涵，经常被人们谈论为建筑的"基本性"和"本质"。这也是建筑学作为一种学术和职业，在社会上的立足之本，所谓建筑的自治（Autonomy），有相当一部分应该源自于此。

随着社会的不断发展，建筑师必然要面对新的社会问题，要以新的方式与方法进行探讨。而这种新的问题和新的方法，并不会削弱建筑学的"主体性"，反而会更加增强建筑学的"主体性"。建筑学科，正是如此这般几千年生生不息地发展过来的，这是一种建筑学的力量。

我们也可以这样理解：一个新的社会问题发生时，正是对我们建筑学术之"主体性"的一种考验。工业建筑与遗产这一新的社会问题，也是一项正在发展的社会事业，需要从建筑学角度来重新合理定位这个课题。以建筑师的角度，我们需要对这一情况有本学科与本职业恰当的态度，并配合相应的工作方法。宏观地讲，作为一种社会发展现象，我们必须理解其作为新的设计问题之由来；作为一种社会新问题，我们也必须有相应的设计策略来应对。

"建筑馆"，正在于通过合理地利用原机械工厂的空间与场所，巧妙地改造成了建筑学院的功能与内容；原厂房的更新，改善了与环境空间的融合，校园环境也因此得到了复兴；在整个过程中我们看到原工业建筑的实体、空间以及器械被大量保存并加以利用；因此，工业遗产的保护工作也被做到了极致。笔者以为，这正反映了一位优秀的建筑师，以其建筑学的信念和理想，坚持建筑学的主体性，不断进取，策略性地实践新的设计方法，从而得以创造性地解决社会的问题。这其中反映的正是建筑设计的智慧之果。

为此，笔者很欣赏"建筑馆"这个名称。将废弃、腐朽之工业"怪物"重新注满人性的活力，我们能感受到的是建筑学的力量。

注释

1.关于"工业遗产"或"产业遗产"的概念以及相关论述，参见：张松、王建国等学者，《建筑学报》2006（8）。

2.参见《下塔吉尔宪章，The Nizhny Tagil Charter for the Industrial Heritage》，方婉丽译，百度文库，http://wenku.baidu.com/view/2b4aa53343323968011c92bb.html.

3.参见：张鹏举，《空间引导功能》，《建筑学报》2010(4)，从校园文化活动中心到建筑馆，作者原出发点是以北京798、上海8号桥等为案例的效仿榜样。但当学校将之改造成为建筑馆，并以此为条件来申请建筑学全国专业教育评估，整个任务就变成了十分务实的建筑教学功能引入的再利用设计了。

4.由于空间不够，内蒙古工业大学建筑工程学院在建筑学院向学校提出改造机械工厂的报告的同时，向学校提出了要建筑学院搬出去的报告。学校领导允许建筑学院来承担此改造任务，则是以建筑学院必须申请建筑学专业评估为条件的。

5."关税联盟"（ZollVerrein），位于德国鲁尔工业区改造计划的核心区的埃森（Essen），整个改造规划与设计云集大量国际著名建筑师、规划师。如库哈斯(Rem Koolhass)、福斯特(Norman Forster)等；"阿尔伯特仓储区"(Albert Dock),是英国利物浦港口区改造的核心项目，也有斯特林(James Stirlin)等著名建筑师参与了规划设计。两处在改造成功后，均成为联合国教科文组织(UNESCO)的"世界文化遗产"。

注：此文摘引自《新建筑》2011.05

Zhao Chen Power of Architecture
——A Discussion on core value of Architecture in the Preservation of Industrial Heritage with the Case of Architectural Hall of the Inner Mongolia University of Technology (IMUT)

Abstract: The preservation and reuse of industrial heritage have been an important content of social development, architecture is automatically involved to contribute professionally, but there still exist some misunderstandings in the profession. The Architectural Hall of the Inner Mongolia University of Technology, as a successful architectural case of redevelopment from original mechanical factory complex, demonstrated that within the redevelopment process, we should not simply concern the preservation of industrial heritage but also emphasize the reuse of industrial building and re- adaptation to urban environment, thereby reflecting the core value of architecture.

Keywords: Industrial Heritage, Misunderstanding, Preservation of Industrial Heritage, Reuse, The Architectural Hall of the Inner Mongolia University of Technology, Core Value of Architecture

The Architectural Hall of the Inner Mongolia University of Technology (hereinafter referred to as "The Architectural Hall") has been rebuilt from an abandoned machinery factory since 2009. The Architectural Hall, which remains the original factory rooms as well as the original image of the factory, has now become a place of teaching and scientific research in Inner Mongolia University of Technology. Viewers are amazed by the fantastic change of the place from a desolate factory into an architectural building for academic purpose. Deeply impressed by the transformation, the author's view of the paper is using the Architectural Hall as a successful case of industrial architecture protection, combining with the social meaning of industrial architectural heritage protection and reuse, to explore the academic significance of such cases.

Cultural heritage protection of industrial architecture is a very popular social undertaking and a public topic, which has been given considerable attention from all walks of life. Those famous cases of protective transformation have gained huge attention from the public. The author has been studying this topic in recent years, observing some typical cases both in China and abroad, and participated in the related practical project research work. Through these studies, the author fully recognized that with the concern and support of the society, the endeavor is developing positively and at the same time being misunderstood. The epistemological and methodological misunderstanding will be harmful to the development of protective transformation, especially for the architects. The author defined the phenomenon as: "myth" of industrial heritage protection, and has already shared the opinions at some related academic meetings.

The Architectural Hall, built under the direction of Professor Zhang Pengju, can be used as a case to prove the author's view of adhering to the body of the architecture and the meaning of architecture. The author hopes that through studying the case of the Architectural Hall to further elaborate the architectural significance of industrial heritage protection.

I.Industrial Heritage Protection as the New Concept of Architecture

First of all, as a new architecture content, the basic concepts of industrial heritage protection and utilization will be discussed. Industrial heritage protection and utilization has been developed for several years in some developed countries, especially in Britain, Germany and other European countries where industrial revolution had been originated. Even in China, industrial heritage protection and utilization has been developed for more than 10 years. Due to the phases of development of human civilization, the world is entering into the post-industrial era, and the era of informatization and urbanization. With the development of the society, the original industrial factories, which once produced a lot of benefit for the society, have been closed, transferred into other usages or moved to other areas, etc. Therefore, many original industrial buildings and land have been changed for other usage, and some even been idle and wasted. How to reuse these industrial buildings and industrial land has become the new topic of architecture. Because of the rapid development of China, the phenomenon is especially abrupt here, and it's hard to cope with architecture alone.

As a new social and historical concept, industrial heritage protection has already had a strict and clear definition accepted by most countries. The Nizhny Tagil Charter, formulated in 2003 by international council on industrial heritage protection under the framework of international historical heritage council, has been widely accepted around the world. The Nizhny Tagil Charter highlights that "vital significance can be found in the structure of industrial buildings, in the art and tools of industrial buildings, in the urban landscape of industrial buildings, as well as in other material and non-material performance." national

Administration of Cultural Heritage, Housing Security Bureau, Ministry of Culture in China have issued related documents such as "Wuxi Suggestions" as a guidance document for the protection of industrial heritage. Compared with the developed countries, China started relatively late in industrial heritage protection, and it is clear that the task is more difficult. The requirement to protect the industrial building heritages put forward by the charter and other documents involved quite a large field. According to the Nizhny Tagil Charter, there is almost no difference in the definition of industrial heritage and other historical and cultural heritage protection. It is easy to take for granted that industrial heritage protection should be equivalent to other historical and cultural heritage protection. So, as an architect, the protectionof industrial heritage buildings has been added to the previously need to protect the historical architecture.

In fact, most of these charters focus on discussing the significance of protecting the industrial heritage, while the architects' problem is how to do it. From this perspective, the charters and related discussions can offer little help for architects. Planning and designing plays an important in architecture, and the contents of the industrial heritage comparing with cultural heritage are quite different in the spatial characteristics. The Nizhny Tagil Charter mentioned: "industrial cultural monuments include buildings and machinery, mining and processing workshop, production workshop, factory site, warehouse, the place for production and transformation, residence, religious worship or education and other social activities associated with industry." These buildings or structures tend to be huge and spacious in scale, and include a lot of machinery and equipment. Such places are difficult to connect with people's normal life, work and other activities. Simply to say, industrial heritage has a machinery rather than humane scale and space. Although industrial architecture as industrial culture heritage should be respected and protected, it's not realistic to equal these great "monsters" of industrial civilization with other cultural heritage.

The Architectural Hall undoubtedly belongs to the cases of industrial heritage protection. The success of the Architectural Hall is due to that the project's main architect Zhang Pengju, the project's decision makers and performers didn't simply take it as a cultural heritage protection. However, there are quite a few architects simply equal cultural heritage protection with the industrial heritage production. The author thinks that this equivalence is actually a misunderstanding.

II. Architectural Misunderstanding of the Industrial Heritage Protection

In the face of the new social phenomenon— industrial heritageprotection, we need to use some new aesthetic value about architecture, which means that it is not necessary to give too much repair or decorations to the rough place of the industrial buildings. It is wise to accept the rough aesthetic perception of the "mechanization". And it has become an important mission for us as architects to realize the activities at a part human scale inspace of the whole industrial scale. Nevertheless, this writer also doubt whether it is the main duties for us architects because what

we can see is that some photographers, designers, journalists or media producers are completely competent enough to make it through some "show field" which usually produce a better result than the tasks completed by the architects.

As an architect, confronting the new social needs, it is a necessity to obtain new ideas and methods. It is certain to bring unavoidable misunderstandings if we continue to use the accustomed ideas and means as used in cultural heritage buildings which simply emphasize on the strict protection in architectural design. From the point of the job nature of architectural design, there are mainly two misunderstandings.

1. Misunderstanding One: Protect It or Take Advantage of It?
It seems that we have talked too much about the industrial heritage protection, while we have forgotten about the utilizing problems connected directly with it. It fact, the two are closely connected to each other and can not be separated from the perspective of architect since it might not exist that there are industrial heritage protections without being associated with utilization. Even though it exists, it has less to do with the architects' work.

For the architects who undertake the planning and design of certain engineering projects, the more important question that they care about compared with the one that they discuss on the importance of the industrial heritage protection in theory, is how to make use of these "heritage" properly. That is: to research, plan and design the recycle of these industrial building spaces. For most of the time, architects need to put new program and function again and arrange new activities properly in the abandoned industrial buildings and environment, which means its design is limited enormously that it has become impossible to regard it as simply as mould of "functionalism" under the function of "form follows".

The Architectural Hall, as an actual engineering project, is just based on the question of reality which is how to make use of the former machinery works of Inner Mongolia University. After being discarded for many years, this machinery works was expected to be torn down by the school to make use of it. At this moment, out of the sensitivity of an architect, Zhang Pengju and his colleagues proposed a reasonable idea which is to remould it instead of demolition, which may help the school to solve the problem of spatial place for teaching and scientific research in Architecture College with comparatively less investment. And after accepting this advice, the school leaders made use of this opportunity to make it into the place remoulding task of the Architecture College, of which the background condition is that Inner Mongolia University of Technology Architect College originally had no its own place, and it shared one building with the Architectural Engineering Institute. The space was very limited and hard to reach the spatial place standard required by the National Architecture Teaching Evaluation.

It is obvious that making use of the discarded machinery works to get the effective spatial place of the Architecture College has become the fundamental purpose of this project. Use was the starting point.

Of course, appropriate use necessarily means successful protection in a certain sense. The significance of Architecture always is to integrate and coordinate all factors, rather than a single factor in a purest sense. The success of the Architectural Hall is just that it combined the use and became the case of industrial heritage protection of Architecture College. Further more, it is also the most sufficient and the most convincingly practical demonstration that to use for the first and give consideration to the protection.

From the nature of occupation to analyze: confronting such social problems as industrial architecture heritage, it should has different division of labor in society from the nature of occupation; to the antiquarians, it should naturally be the questions of protection, which is also the basic value guidance in the social culture circles; while to the architects who carry out the planning and design tasks, it is a question focusing on use. Of course, we hope that it is the use with protection. However, use goes first. When protection and use are in conflict and we can not have both, part of it must be torn down. It's impossible and not realistic to protect the industrial heritage completely as the historical and cultural relics from the ancient remains. Especially to those industrial buildings and structures, if we still overemphasize the protection, it would lead to the unreasonable development of the society. The research and reflection on these questions are just characterized by the architectural design and planning tasks. Reflecting from the social occupation and specialization of work perspective, lacking of the architects' focused reflection on the recycle problem will surely lead to the bad result of the whole industrial heritage.

Lots of architects forget their professional duty and quality, when facing the tasks of industrial heritage, and just talk highly about protection with the heritage scholars and the cultural scholars, from which they seem possess the trend of the times and the social prevailing custom without realizing that the architecture value reduces or loses with it. What's worse, our architecture academic world is just like this. They only emphasize on protection and educate the young generation of architects in this way. When we incite our architects to be the one as the social cultural scholars who only pay close attention to protecting the industrial heritage and do not know how to make use of the industrial architectural spatial place, such kind of "protection" surely will fail, from the society perspective, when the architects only speak highly of the protection and do not obtain the competence of use.

It is gratifying that the case of the Architectural Hall focused on use. This writer is happier about that the Architectural Hall as the College of Architecture engages the students of Inner Mongolia University of Technology to realize the reasonable use of this industrial factory building in experiencing and enable them to understand correctly the significance of industrial heritage at present and in the field of architecture. In fact, there are many such kind of successful cases at home and abroad whose focuses are all use, such as the world famous cases: ZollVerrein, in Ruhr Industrial Base, Germany; Albert Dock, at Liverpool port, UK; etc. are all the successful remoulding cases whose focuses are all use. And their cultural heritage significance is shown after the successful remoulding. While some so called Chinese famous cases were just completed by some sensitive artists and designers who voluntarily made use of the old workshops and warehouses appropriately which was approaching death, and turned them into work rooms, offices, and so on before the scholars and authorities have realized the significance of industrial heritage protection. For example, both "798" in Beijing and the bank along "Suzhou River" in Shanghai are not the achievement of advanced protection theories, but the product of practical use.

2. A Misunderstanding in Architecture toward the Industrial Heritage Protection

We also seem to talk too much about the industrial problems, and forget the problems of the city or the environment, which is the second misunderstanding.

Most of the industrial building or heritage is associated with a larger range of urban environment, we will inevitably encounter the so-called the concepts of "protection" and "revival" when we are planning to protect and use an industrial area. However, when the region occupied by an industrial building is concerned, are the protection and rehabilitation in the end for the industry, or for the city or the environment? In most cases, the problem has been mixed up. I have participated in a design comment with the theme of International Work Camp; the author put forward the proposition of "industrial rehabilitation", which precisely misled the design and study of the participants. Some problems in his works reflect more clearly the misunderstanding and some designs extremely care symbolic elements of industry building, regardless of the relation the of the whole plant with the various parts of the city, especially with the transportation and streamline. The particularity of the rigid system and land management systems of state-owned enterprises China is making a lot of enterprises in the industrial plant area create "Creative Industry Park" and stuck in their closed range, regardless of the city's surrounding. From the perspective of urban development and environmental issues, it is actually very harmful. This situation has been reflected in many projects in recent years, even in the so-called famous project. Such as "1865 Creative Industry Parks" of previous Nanjing Chenguang machinery factory and Jinling machinery manufacturing Bureau which have developed much since 2003 without enough consideration of the relationship with the city, and bringing no attention of the relevant official departments. Once the projects have further development, the problems of the city's transportation, landscape, format and others are bound to continue to occur and are difficult to solve, which leads to over emphasis on industrial heritage, ignoring the bad consequences caused by city revival and the attitude to evade the crucial point in the academic circles for a long time.

Even the famous Beijing "798", in the author's view and in the sense of urban renewal, can not be considered a success with inconvenience of traffic flow, the poor urban spatial location, and implying a considerable number of urban public and open problems.

We should realize that the necessity of the protection of industry itself should be limited, and there is no need to "be revived". From a social point of view, the industry is dying out or moving, which is the law of social development. The real need for rehabilitation and protection of the objects should be the cities, or the environments. In the city, how to adapt to the urban environment after the revival of the original industrial construction areas? The industrial areas are originally a part of the city as well as the important economic zones. The industrial production and industries in the process of social development decline or transfer, which means, from the perspective of the city, re-urbanization of the original industrial regions.

The Architectural Hall, mainly show how to reintegrate the old, harsh and rough plant incompatible with the environment into the present campus. We are pleased to see the interface of the Architectural Hall associated in the campus makes the design of outer space positive. In the courtyard of the Architectural Hall, there are doors, bridges and railings renovated by original factory facilities or components as well as the additional external space facilities such as outdoor venues, landscape walls and seats, creating an extension open to campus. These measures make the project of the building space not stay in the interior itself, but cooperate with the whole campus environment. In the next phase of the expansion project, we can see the revival scene fully coordinated with the development of campus environment. The attitude, taken in the Architectural Hall with positive initiative to match the surrounding environment in space, is that ignored by a large number of designers for industrial heritage protection and transformation, who have not enough care for interaction between cities and the environments.

The environment of the cities with industrial heritages can achieve a true rehabilitation, only when we have a full consideration, in planning and designing projects, of mutual penetration in space, interaction, communication and integration, especially full introduction of city public space into the original industrial areas.

III. Reconsider the Industrial Building Heritage from the Subjectivity of Architecture

The comprehensive characteristics of architecture and profession will be expanded in the field of study when it faces new social problems each time. This means that architects should understand and learn something new, including new theories, new designs, new styles, and so on. And these new content often includes the characteristics of other disciplines, professional, the field, so it seems that we must be involved in other areas, or to be affected by other disciplines. But if we only focus on the meaning of the things that we have been developed, and forget the essence of our own subject and profession, it is easy to lose its own subjectivity. Architecture must face these issues all the time, and the definition of the so-called architecture would be an eternal question. Up from the definition of logical meaning, as connotation and extension of architecture, its concept has the following characteristics: a relatively large uncertainty of the extension, which will lead to the architecture constantly sorbs new contents; but its own connotation is relatively stable, namely, architecture should to solve the social problems creatively regardless of how many new things to face. The relative stability of connotation is often talked as the "base" and "essence" of architecture. It also makes the architecture, as the academic research and profession, keep a foothold in the community. A considerable part of the so-called architectural autonomy should be derived from this.

With the continuous development of society, the architect must face the new social problems and explore new ways and methods. This new problem and new methods will not weaken the "subjectivity" of architecture; instead, it will enhance the "subjectivity" of architecture. Architecture Engineering has developed over thousands of years of life and growth in this style, which is a kind of power of architecture.

We can also understand it like this: when a new social problem occurs, it is a test of the "subjectivity" of our architecture. The new social problems of Industrial architecture and heritage, which is a development of social undertakings, need to be relocating reasonably the position of its subject from the perspective of the construction. With the architect's perspective, we need to have a proper attitude towards this situation, and to work with it. Macroscopically speaking, as a social development phenomenon, we must understand its origin as a new design problem; as a new social problem, we must also have the corresponding design strategies to deal with it.

"The Architectural Hall", through the rational use of space of the original mechanical factory, skillfully transmit the function and content of the school of architecture; the update of the original plant fuses the surrounding environment better, and campus environment gets a revival in the whole process; it sees that the original industrial construction entity space and equipment was preserved and reused in a large amount: all in all, the protection of industrial heritage has been done supremely well. This reflects that a good architect, with his architecture beliefs and ideals, insists on the subjectivity of Architecture while forging ahead, strategically practices the new design method and creatively solve social problems. The wisdom of building design shows out completely.

To this end, I appreciate "The Architectural Hall". When the waste and decay of the industrial "monster" was refilled with the humanity vitality, we can feel the magic power of architecture.

Notes

1. The Concept of "Industrial Heritage" or "Business Heritage" and the Related Discussion refer to Zhang Song, Wang Jianguo. Architecture Journal, 2006(8)

2. Fang Wanli. The Nizhny Tagil Charter for the Industrial Heritage Nizhny Tagil Charter. http://wenku.baidu.com/view/2b4aa53343323968011c92bb.html

3. Zhang Pengju, Function Follows Space and Architecture Journal 2010(4). From the campus cultural activities center to the Architecture Museum, the author originally intended to take Beijing 798, Shanghai No. 8 Bridge as the example. But when the leader asked to transform it into a museum, and to apply for the construction of the national professional education assessment, the whole task has become a pragmatically remodeling design of function introduction following building teaching.

4. Due to the limited space, the College of Architecture of Inner Mongolia University of Technology reported the transformation plan of machinery factory, at the same time, report to move out. University leaders allowed the College of Architecture undertake the remodeling task, but should apply it for the construction of the College of Architecture Professional Assessment for the conditions.

5. "Customs union" (ZollVerrein) locates in Essen, the core area of the Ruhr Industrial Area renovation plan. A large number of international famous architects and planners join in the entire renovation planning and design gathered, such as Koolhaas (Rem Koolhass) and Forster (Norman Forster); "Albert storage area" (Albert Dock) is the core retransformation project of the port in Liverpool District, Stryn (James Stirlin) and other famous architects involved in planning and design. After the successful remodeling, now these two subjects are the world's cultural heritage of UNESCO.

Note: It quoted from *New Architecture* in May, 2011.

王兴田 "逆设计"
——内蒙古工业大学"建筑馆"改造设计评述

美丽的童话"灰姑娘和水晶鞋"的故事,让每个人的童年都留下了美好的记忆:王子带着灰姑娘留下的小巧精致的水晶鞋,满城找他的心上人,"哪位姑娘能穿上这只水晶鞋,我就娶她做我的妻子!"城里许多年轻漂亮的姑娘都争先恐后地试穿那只鞋,却都不合适,最后只有当灰姑娘把脚伸进水晶鞋中的时候才发现,这鞋就是专为她的脚而铸成的,从而成就了王子与灰姑娘美满幸福的一生。

以试穿水晶鞋来寻找梦寐以求的心上人,与我们建筑设计中的改造设计有得一比。六十年代建造的铸造车间,谁也没想到,时隔四十余年找到了他的"灰姑娘",那新旧时空的穿越结合让老厂房焕发出了新的生命!

一般来说建筑设计是按功能需求将场地及环境等因素进行综合分析和解答,从而给出适当的空间分区并组织动线,简言之,其设计逻辑是先因后果的思考过程。的确这种符合逻辑的"顺"设计显得合理、谐调,也具有美感,但却可能流于平庸。而"逆"思维所设计出来的建筑看起来虽然逻辑感不强,也许很多地方还处在矛盾中,却在矛盾的博弈中获得"顺"思维达不到的效果,呈现出耐人寻味的思考轨迹和深度。历史留下的工业遗迹,由于原始的工业技术已被淘汰,空间丧失了其本来的使用意义,留下的只是一个建筑躯壳,似乎已成为一堆荡然无味的垃圾。建筑师在接受业主(内蒙古工业大学)委托时被要求在这个铸造车间所在的基地上,建一座新的建筑馆。然而建筑师经过思考,从建筑学教育以重交流、重体验、重实践为特点出发认为不如将新内容注入老空间,这样既能留住"文革"年代特定的历史记忆,又能让它未来的主人——一批批建筑学的学生们在这里体验经历沧桑的"新"建筑。就好比将那六十年代不起眼的旧车间擦亮成一双水晶鞋,为它寻找到属于它的"灰姑娘"。

为此,建筑师逆势而上,将因果倒置,开始了改造设计。然而"逆"设计过程中最令人烦恼的往往是矛盾和结果明知存在却不能够准确地预见和把握,而在施工过程中这些问题接踵而至,似乎这时设计才真正地进入状态。也正是随着许多难以预料的问题被逐一解决甚至突破,设计才由被动变为主动,营造出顺向思维设计中达不到的空间效果,成就了意想不到的魅力与特点。整个施工就这样在"逆"设计动态中不断推进。因建筑由高大单层空间内加改造成三层,人在三层上会非常接近地观看原有的大跨混凝土梁,近距离感受到它的巨大尺度。这种因距离而改变了的尺度,看似不合逻辑却让身临其境的人受到了强烈的震撼。同时,站在三层上也能近距离地体会天窗等构件的细节,让人细细地品味着那个年代的痕迹。

同样在内外空间接口的开窗设计上,为原汁原味地保持原建筑的外立面整体效果,改造的内部空间的开窗往往并非合理行事,有些只能无奈地就位。但正是这样的无奈,却带来了别具一格的感受。原有的天窗被保留着,这不仅让空间深处的采光得以解决,还给开放流动的空间洒满光影的魅力。随着时光变幻,仿佛空间跟着转换,心情也在舞动。透过办公空间内低于视线的落地窗,当人站立时能俯瞰院落近处的景象;当坐下来学习工作时,校园内的远景又尽收眼底,这样的效果仿佛是设计有意而为。

顺着东边的楼梯拾阶而上便看到楼梯平台上的T形玻璃面,正好与下一层的沙龙空间形成视觉交流,别具匠心。殊不知这原是拆掉穿越楼板的旧设备留下的T形洞口,设计者用最简单的方式覆上安全玻璃,并保留旁边两台大型设备作为艺术装置,与周围的建筑空间浑然一体,唤起了人们对工业时代的记忆……

不同标高的开放和半开放空间通过楼梯、连廊连接起来,而相对封闭的教室以透明或半透明的隔断来保持其相对安静的空间流动性，最终所有空间都向中庭展开。共享中庭位于L形主体建筑转折处的最佳位置，它把所有子空间汇集了起来。在共享中庭一侧的地面上，有节奏地排列着几个木制的休息台座，它们也是由设备拆除后留下的基座。其侧面地板上还保留着原来用作通风的井口，它们是原始的被动式"地源热泵"，经实践在夏季也能使共享厅内温度降低2~3℃。进入门厅前首先要分别从两条小路穿过由建筑自然围合成的院落，其间保留着原有的大树和用设备改造成的景观艺术装置，这一过程好似形成了前奏，张弛有序并收合启承，进入共享中庭时，让尺度并不大的中庭空间显得轻松舒适。加之室内外材料的一致性，感觉空间在不知不觉中延伸、转换……

为了取得更好的抗震效果，在支撑吊车的梁的牛腿上加焊了从设备上拆下的型钢来加固节点，这种做法，使混凝土与钢构如期相会却并不"有机"结合，既在情理之中又在预料之外。

建筑馆的空间价值是通过其内容、形式与空间的高度统一来体现的。在充满活力的空间中，更因有主人——建筑学的学生们的参与而显得更具意味。建筑学的最佳教育方式是老师与学生，包括不同年级的学生之间的交流互动，从而获取不可言传的学识。如此，空间开放、流动的建筑馆无疑给学生们创造了极好的交流氛围，从不定性的空间到多接口的细部，从多样的材料到赤裸的节点，无不预示着学生们将在其间获得潜移默化、无师自通的建筑领悟。

旧建筑的改造再利用虽已不是新鲜话题了，但在我国各城市对旧建筑改造再利用的设计和实践中还是存在不少误区，或借用保护传统的名义搞急功近利的商业操作，或过度改造而不能准确还原建筑本体的价值，抑或是新功能、新内容与原有建筑空间格格不入等等。

在本设计中建筑师体会到了旧建筑的改造设计比新建筑设计更具有难度和挑战，是一种"逆"向思维方式设计，其过程是痛并快乐着的。设计遵循的原则是对旧建筑的足够尊重，借助当代技术手段从那个时代的思考方式出发来处理现实问题，顺形就势，自然而然地将当代所需的内容植入其原本的面目中，使旧建筑发挥出当今社会应具有的物质价值和精神价值，使徜徉于其间的人们在满足需求的同时，能深切地体会到那个时代的文化印记，感受到历史的长河在我们身上流淌。这才是旧建筑改造再利用的真正意义！

注：此文摘引自《新建筑》2011.05

Wang Xingtian Reversing Design
—— The Remark on the Reconstruction of the Architectural Hall of the Inner Mongolia University of Technology

In everyone's childhood memory there is a romantic fairy tale: "Cinderella and the Crystal Shoe". The prince looks for his sweetheart all around the city with the crystal shoe in his hand, "The girl who can fit into the shoe is the one I want to spend my rest life with!" When they get the news, many beautiful girls in the city are afraid of missing the opportunity and go immediately to try the shoe, but none of them is the right person. Finally, when Cinderella puts her foot into the crystal shoe, and it fits her perfectly. It seems that the shoe is made for her particularly. Thanks to the crystal shoe, the story ends in a happy wedding.

This story is analogous to the reconstruction architectural design project. The casting foundry built in 1960s, out of everyone's expectation, found his "Cinderella" after 40 years. The combination between the past and the present vitalized the old workshop.

Generally speaking, architectural design is to analyze and settle down the relationships among the site, environment and other elements comprehensively according to their different functions, in order to leave proper spacial division and organize moving line. Simply speaking, the logic in the design is the cause first and then the effect. This mode of design is reasonable, coherent with sense of beauty, however it might be mediocre. While the building designed under the reverse thinking mode is not highly logical and many aspects of it are in contradiction, while unexpected effect is achieved in the confronting between the contradictory elements, and unique thinking trajectory and depth are presented in such design. The industrial relic left in history has lost their original functions; only the outer form is left behind. It seems that they become useless garbage. The architect was requited to build a new pavilion on the spot of the old casting foundry when he accepted the proprietor (IMUT)'s commission. While after consideration, the architect thought it is better to infuse new content to old space according to the communication-experience-practice oriented feature of architectural education. Thus, on one hand, it can keep the historic memory of "cultural revolution", on the other hand, its future owner— batches of students major in architecture can personally witness the "new building" with the turbulent past. It is just like polishing the old workshop and transforms it into a shining crystal shoe and finds him his "Cinderella".

Therefore, the architect reversed the order of cause and effect and began the project of the reconstruction design. While in the process of reversing design, what troubled the architect most was that he knew the contradictions would happen but how they would happen was unpredictable and uncontrollable. In construction, the problems came one after another. It seemed that the design was really getting into the groove then. When these unpredictable problems were worked out even solved one after another, this project is turning into the right track. The special spacial effect that is unexpected in the straightforward design mode is created in the reversing design and unimaginable fascinating effect is achieved.

The constructing is making constant progress in the "reversing" design mode. Because the inside of the building is transformed into a three-story-space, people standing on the third-floor can see the concrete beam and feel the boldness of the design in a very close distance. This effect of being in the space personally is created by changing of the distance. It leaves very deep impact on everyone's heart. At the same time, standing on the third floor, people can see the details of the skylight and other components very closely so that they can relish the remains of history. And in the design of the window cutting in the interface of the interior and exterior space, in order to keep the overall original taste and flavor of the facade of the building, the transformation of the interior space window cutting is not a reasonable design, and some of it is just put into position because of having no other choices. However, this resignation created unique effect. The old skylight was remained to get enough sunshine and the flowing of the light and shadow in the open space is very fascinating. As time is going by, space is changing with it, and the heart is dancing along. Through the French window that is lower than the sight line in the office, people can see the view in the yard nearby. When they sit down to study or work, the panoramic view of the campus will be held. It seems that this effect is achieved on purpose in the design.

Up along the stairs in the east of the building, people can see the T-shaped glass surface on the stair platform, they can have a visual communication with the down stair salon, thanks to the ingenious design. Whereas, this T-shaped space is left behind in the pulling down of the old floor, and the architect designs to cover it with the shatter-proof glass. Two large pieces of equipment are retained there as the artistic decoration and they become the integrated part of the surrounding building reminding people the memory of the industrial age...

The open and semi-open spaces of different elevation are connected by stairs and corridors. Transparent or semi-transparent partitions are built to make sure the quietness of the classrooms and the fluidity of the space. And all the spaces are stretched toward the center. The shared atrium space is put in the best position of the flexed part in the L-shaped main building. All of the sub-spaces are bound. On the ground of one side of the shared atrium space, several wooden pedestals for resting are placed in rhythmic pattern. They are actually the foundation beds left after the equipments are pulled down. On the side floor of it, the well head for ventilation is retained. They are the raw-passive-ground source heat pump, which can lower the temperature of 2~3℃ in the shared atrium space. To enter the building way, people have to walk through the yard formed naturally by the enclosed building. Within the yard, the old trees and the artistic installations transformed by the former equipment are retained. This part of the building is like the prelude in the music, it is well-aligned and naturally- connected. Due to this design, the shared atrium space which is not quite spacious seems to be relaxing and comfortable. The consistency between the interior and outdoor material makes the space more stretching and expanding.

In order to get better effect of seismic performance, the structural steel pulled from the old equipment is welded on bracket supporting the crane. So that the concrete and the steel structure are united but are not combined organically. It makes sense but it is out of people's expectation.

The building's spatial value is reflected in the high unity of content, form and height. In the space full of vitality, the owner of it—students major in architecture make it more lively. The best teaching methodology of architecture is the communication between teachers and students of different grades, so that the knowledge that can not be expressed in words can pass down from generation to generation. Thus, the building with open and fluid space creates very good communicative atmosphere, from the indefinite space to the multi-interface details, from various of materials to the naked joint, the students can comprehend a lot about the art of architecture from the building unconsciously without the teacher.

The reconstruction of old buildings is a not a new phenomenon. However, in many cities in China, there are shortcomings in the design and practice of the old building reconstruction. Some of them seek instant commercial benefit under the excuse of tradition protection, some of them can not restore the original value of the building and sometimes the new function of it can not match with the space of the old building.

In this design, the architect discovers that the old building reconstruction is more difficult and challenging then new building design. It follows a "reverse" thinking mode, and the process is painful and happy as well. The principle this design keeping up to is to hold enough respect to the old building. With the help of contemporary technological means and the thinking mode of the old time, practical problems can be solved. The content that is needed in contemporary is implanted into the original look of the old building, so that the material and spiritual value of the old building can play their own parts in the present society. People's needs are satisfied when they are wandering in it, and they can feel the cultural marks of the old time and sense the river of history flowing into the present day. This is the significance of the old building reconstruction.

Note: It quoted from *New Architecture* in May, 2011.

黄居正　空间：记忆的装置

在《密斯评传》(Mies Van der Rohe: A Critical Biography, by Franz Schulze)一书中，弗兰茨·舒尔茨讲述了一则现代建筑史上的轶闻：1921年，密斯在柏林开设的事务所与表现主义建筑师雨果·海林(Hugo Haring)曾共享一室，两人常常因为空间观念的大相径庭发生争执。海林认为，建筑中的每个功能空间应该拥有与之相应的独特形态，而密斯却认为"空间应该足够的大，大到可以自由地徜徉其间，而且是一种均质性的空间……功能是模糊不定的，况且会随着时间的变化而变化，固定不变的功能并不存在。建筑永远比功能恒久。"

抛却对现代主义的偏见，从某种意义上而言，密斯是对的。建筑可能存在十年、百年，乃至历千年之久，而原初设定的功能早已不再，若建筑物内部的空间越是开放、通透，重置功能的可能性就愈大。譬如工业时代各城市中大量遗留下来的厂房，因产业转型，被置换成艺术展示、SOHO、商业等新的功能，在近几十年间，出现了德国鲁尔工业区、泰特美术馆、北京798、上海M50等相当多改造与再利用的成功案例。

内蒙古工业大学建筑馆，是由原属内蒙古工业大学的机械厂房改造而成。改造的第一步，是清理旧有空间中的赘余之物，保留所有能令人联想和唤起工厂意向的特征，尤其是那些朴素、真实的具有工业时代特征的要素。用型钢补强了某些薄弱部位的钢筋混凝土框架柱和屋顶构架被刻意地裸露了出来，其强烈的视觉特征暗示了建筑的原初属性，同时，高达十几米空旷的室内也为建筑师提供了空间操作的丰富可能性。

在置入建筑学院所必需的新功能时，建筑师对不同要求的各类教学、活动空间做了适当的区分。在主入口的西侧，通过加层，图书馆和评图室设置在南向的一、二层，完全开放的美术教室则被置于顶层，沐浴在高侧窗洒入的均匀光线之中。我猜想，长期浸淫于教学，并时任学院院长的鹏举兄，自然对展开教学所需的空间条件熟稔于心，为了打破司空见惯的平庸空间，于是，美术教室不设隔断，向四周开放，就像一个空中舞台，在此展示出一幕幕空间戏剧。教师、学生参演其中，严肃的学院气氛瞬间消融，像音乐中出现一个悦耳的强音，空间似乎在这儿跳跃了起来，呈现出欢快的色彩。在北向，则植入了展览空间，通过台阶联系各个不同的高程，这些高高低低四个不同高程的不平整楼板，与其说是功能性的，不如说是表现性的，或者说是出于空间形式的需要，它们与那些质朴的红砖砌筑的展示墙一起，构成了一个连续不断的空间，这一连续性的空间却又与密斯式的透明和流动相悖，更强调在行进中体验间歇和停顿，从而让每个空间保存其自身存在的理由。

主入口的东侧，安置了功能性较强的专业教室、视觉实验室、行政办公等空间，通过U玻等半透明材料的使用，使空间围而不合，充分地保持建筑整体的开放性。在转角的三层通高空间内，放置了一步直跑楼梯，直抵顶层的教室。这道冷峻的钢构楼梯，插入此处高扬的垂拔空间，颇似意大利艺术家卢西奥·丰塔纳(Lucio Fontana)"割破的画布"所产生的效果，拓殖了空间的另一重维度。

在建筑之内蜿蜒回转的其他一些楼梯和桥，同样被建筑师赋予重要的作用，它们不仅承担着交通的功能，连接建筑中的各功能区块，更重要的是，它们是整个建筑空间中一个个欢腾的音符，活跃着空间的氛围，丰富了空间的色彩。

列维·斯特劳斯在《忧郁的热带》中曾提到："有一次，一个巴西女学生第一次到法国之后，泪眼汪汪地跑来看我，巴黎那些因岁月而阴暗的建筑物，她觉得很脏。她判断城镇的唯一标准，是看它有多白、多干净。但是那些面对纪念性建筑物所引起的超越时间的沉思，那些最漂亮的城市所具有的亘古常新的生命，那些不仅仅是单纯为了满足都市生活功能，而且是沉思与回想对象的建筑物，都是美洲城镇所不具备的性质。"

城市如此，建筑也如是。工业遗产凝缩着刚逝去不久的一个时代的集体记忆，保留具体的物质形态和时间印记，是保护和再利用的题中应有之义。因此，原厂房中的机器装置，如艺术沙龙与办公区中的冲天炉、模型室中的大型沙漏、庭院中的煅烧炉、草坪中的重型机械，它们或被保留，或改造成特殊的构件，进行重新的维护和刷漆，并与空间中引入的那些新元素一道，加强了建筑学院曾经作为工厂的身份，它们让光阴具备了形态，让时间富有了深度，让记忆锚固在具体的场景之中。

但设计的策略没有仅仅停留在对于过去的浪漫再生，而是展现了学习、交流和娱乐之间的辩证认识，新生成的空间变身为激发同学们敏感、自由、想象和欲望的媒介。

注：此文摘引自《平实建造》

Huang Juzheng　Space: Mechanism of Memory

In *Mies Van der Rohe: A Critical Biography*, Franz Schulze told an anecdote in the modern architectural history. In 1921, in his firm in Berlin, Mies Van der Rohe shared one room with Hugo Haring, an architect of expressionism. They would often quarrel because of their drastically different concept view of the space. In Hugo Haring's view, each function area in a building should have its own corresponding form. But Mies Van der Rohe thinks differently. In his opinion, the space should be large enough for people to stroll freely. The space should also have unified characteristics with vague distinction of functions. With the time passing by, the functions should change accordingly. There does not exist a fixed function and the building always lasts longer than the function.

To some extent, Mies Van der Rohe's idea is right if the prejudice against the modernism is abandoned. A building may last ten, a hundred and even a thousand years. After so many years, the previously designed functions may have already disappeared. The more open the inner space of a building is for re-modelling, the more likely it is for the redesign of the functions to occur. For example, due to the industrial repositioning, a lot of old factories of the industrial age have been renovated for the new functions of art exhibition, SOHU, and business center. In recent years, such successful transformation has occurred in Ruhr Industrial Base in Germany, Tate Gallery, Dashanzi Art District in Beijing and M50 in Shanghai, in which renovation and re-modeling have been completed.

Architectural Hall of Inner Mongolia University of Technology was built on the foundation of the old factory of Inner Mongolia University of Technology. The first step of this transformation work is to remove the remains in the old factory so that the factory can be evacuated for further renovation. Some of the elements of the factory are preserved to remind people of the industrial age, especially the plain and authentic elements representing the industrial age. The pillars made of the reinforced concrete are used to strengthen the fragile sections. Structural steel, which is exposed in the hall, is used to fortify these pillars and the structure of the roof. This powerful visual impact indicates the original function off the building. At the same time, the expansive interior space with a 10 - meter high reaching roof gives the architect more flexibility and freedom for his creative design.

In the process of instilling different functions into the Architectural Hall, the architect divided the space into different zones of functions such as teaching and other activities. At the west side of the main entrance, the space is divided into two floors. The library and the Drawing Review Room is located at the first and the second floor in the south. The open fine art classroom at the top floor is illuminated by the smooth sunshine coming from the top windows. I can imagine due to his devotion to the teaching work, as the dean of Architecture College, Mr. Zhang Pengju has remembered all the detailed room requirements for the teaching work. In order to abandon some mediocre traditions and bring some fresh elements into the space, Mr. Zhang made the fine art classroom an open space like a stage in the air, where the students and teachers can participate in all kinds of performances. In this way, the serious academic atmosphere has suddenly disappeared. The space becomes lively with the bright and pleasant colors. In the north, the exhibition space is instilled. The different levels of the steps are used to connect different floors. The four-level space is more of the expressional purpose than of the functional purpose. For the sake of the special structure of the space, this four-level flor, together with the walls built with the red bricks, forms a continuous space, which is contrary to Mies Van der Rohe's concept of transparency and fluidity. The emphasis is put on the interruption and pause in the movement, which enables each room to have its own reason of existence.

The classroom for specialty courses, visual laboratory and administration offices are located at the east side of the entrance. The use of such semi-transparent materials as U glasses makes this space a continuous and open area, which helps to keep the overall open characteristic of the building. At the corner where there is an open area with the height of three floors, a steep staircase is installed reaching the classroom at the top level. This solemn looking staircase, which stands here like a tower, reminds people of the effect of "The Cut Canvas" created by Lucio Fontana, an Italian artist. In this way, another dimension of the space is created.

Staircases and bridges winding through the building are also used by the architect to give them more functions. They not only help to connect different function zones in the building, but also act as a lively music note in the building, which can create a lively and colorful atmosphere.

In *Tristes Tropiques*, Claude Lévi-Strauss once mentioned that "Once a Brazilian girl student went to Paris for the first time. One day, she came to me with eyes full of tears because she could not stand the time-honored buildings in Paris. The dark colors of the buildings, in her opinion, are dirty. Her only criterion for judging a city is white color and cleanness. The meditations resulting from observing the memorial buildings, the eternal life of the most beautiful cities with a long history, the new functions of the buildings not for meeting the needs of the urban life, but rather for reflection and meditation are the unique features which an American town does not have."

The same is true for the buildings. The industrial remains are a mark of an age which has just disappeared. The physical property and the mark of the age are preserved in order to protect and reuse this old building. Some of the machines and facilities in this old factory are kept intact. Others are transformed into unique building materials through the work of repair and painting. Some of the examples include the cupola furnace in the art saloon and office zone, the large sand glass in the model room, the calciner in the courtyard, and the heavy machinery on the lawn. Together with the newly introduced elements, these old machines indicate to the visitors the former identity of the Architecture Hall, factory. They enable the visitors to have a better understanding of the past, the specific events in the history and the special memory in the particular scene.

The strategy of design, however, does not stay in the romantic past. Rather, it represents a new understanding of the relation between study, communication and entertainment. The newly created space has become a medium to stimulate the students' sensitivity, freedom, imagination and desire.

Note: It quoted from *Genuine Constructing*.

高 旭 从表皮到表情

摘要： 内蒙古工业大学建筑馆扩建，是在一种有着强烈性格指向的环境中完成的。建筑师以理性的方法来应对感性的空间表情问题。在排除了沿用老馆材料表皮的前提下，避免形式表现，借助"砌筑"这一建造方式使新、老建筑产生了共同的性格基础，进而，选择混凝土砌体结构，以诚实的建造逻辑，通过单元布局后的一系列分离、过渡、开放以及动线的流通，使新老建筑之间在厚重、真实的表情中相互对话，相得益彰。

关键词： 建造馆；扩建；砌筑；表皮；表情

一、背景解读

扩建、续接建筑，无论涉及功能还是形式，核心都是关系问题。内蒙古工业大学建筑馆在使用四年之后获得了扩建的机会，扩建目的是将学院四个本科专业的所有设计教室纳入其中，以方便教学与管理。对于这个项目，功能是确定的，场地是确定的，不确定的是，新馆将以怎样的关系呈现在既有老馆所营造的环境中？而新馆的空间表情又将如何呈现？这将是扩建设计中的重点，亦是难点所在。

内蒙古工业大学建筑馆扩建是由老馆建筑师张鹏举教授继续完成的。老馆于2009年由一组位于校园内的旧工业厂房改建而成[①]。该馆一经建成即是自带年龄的"老者"，其空间清晰地携带了旧工业建筑的历史记忆，这也是原馆改建工程设计的一个重要主题。然而，在扩建新馆的设计中，作为决定老馆主体品貌之一的材料——红砖被制度性限制，这在策略上无疑等同于关上了一扇直接有效的大门。建筑师只能选择所谓"对比"的方法，但这扇常被称为"现代"的手法之门却敞口很大，如，在性格上，可以轻盈对比沉稳，也可用简明对比繁复；在材料上，还可用钢构对比砖筑等等，同时，对比域度更是十分宽广。

二、策略解读

事实上，这种对比在校园的尺度中早已存在，在此无须过多纠结，更何况轻盈的钢构亦不是造价有限的教学楼的最佳选择。因此建筑师将设计的关键点确定为空间性格的同一性。在关上了延续相同材料的手法之门后，这种同一性自然诉诸空间表情的表达。

新的空间基于新的功能，倘若用红砖砌就，恐怕也只是出于表皮的视觉策略，而表情是这种表皮背后的性格呈现。一个清晰的道理是，不一样的表皮可以具有同样的表情。换句话说，体验者在同样表情的空间中行进，会自然而然忽略表皮材质的改变。

那么，建造新馆要延续的表情是什么呢？建筑师给出的答案是：厚重和真实。厚重来自于老馆的材质和历史感。真实则主要来自于其自身结构与表皮材料的直接对应。于是，在策略上这两种性格均指向同一种建造手段，即砌筑。新馆的建造选择了混凝土砌块砌墙。它既是结构材料，也是表皮最终的饰面材料。

为了应对砌体结构的空间逻辑，新馆的平面布置采用自下而上的标准单元。在均质分布的六个单元中，五个是功能教室，一个为辅助空间。各单元拉开间隙，在南北对峙的四个间隙空间中，北端两个设置楼梯，南端则开敞面向室外，视线连通该处的老树并增加了休息交流空间，打破走道空间的单调。这既是老馆空间的重要设计策略，亦是形成同样空间性格的重要前提。在首层，处在连接老馆部位的功能单元，将非结构墙体清空后形成展览空间，在人流必经的动线中看展览也是老馆另一做法的延续。

老馆内部空间丰富且开放，而在新馆中由于功能要求和结构体系的限制，空间则略显单调，对此建筑师把关注点放在了二者的连接部分，并把注意力放在了室外空间上。在接通新老馆的通道两侧各设一个小庭园：西侧稍显封闭，成为所有庭院中最安静的一处；东侧则适度开放，融入了校园。它们共同成为专业教学交流空间的延伸，这与老馆的几处庭院在性质上同源。功用如此，其性格更是如此。

同时，这一适度分离，过渡了新馆与老馆在体量上的悬殊，成为尺度上转换的中介。为强化这种过渡的有效性，新馆在西北向增加了入口，向外延伸出一个一层的门厅。此门厅用红砖砌筑，并与老馆围合出一个适于原尺度的更小的庭院。庭院内同老馆一样设流水，增添静谧的气氛。符合并进一步延续了老馆的性格。

在新馆接续中所采取的任何一种策略都不应是孤立存在的，就主体材料和结构的选择而言，本身还有建筑造价的考虑，建筑师所选择的砌体结构，相较于钢筋混凝土结构，经济成本大大降低，亦暗合老馆低造价的重要特征。

性格体验虽是从视觉到心理的过程，但多种感官的整体感受也十分重要，如空气流动带来的清爽感有助于强化从眼睛到心理的转换过程，尤其是在老馆中体验了被动式手法促成的空气流通后，其作用更是不容忽视。试想，在新馆中若不能延续这种清爽，则一切感觉都将无从谈起。于是在新馆的设计中，除了空间表情的表达以外，在空间被动通风的生态策略上也做了有益尝试。平面中各单元之间的缝隙空间不仅是信息交流、沟通内外的场所，同时也是组织通风的重要空间。建筑师在基于效率的走道式空间阻断了穿堂风的前提下，设计了垂直通风的方式。做法是利用单元间的缝隙，通过热压组织气流，并各层独立设置气流路径，避免中和面以下空间的反流。

三、结语

综上，新建筑馆的接续沿承了老馆质朴平和的性格，其空间与材料、建造等相互作用所呈现出的诚实表情，更使老馆与新馆相得益彰。作为建筑学专业教学的场馆，其空间、环境、场所、结构、材料等建筑要素，应会对身在其中的学生产生潜移默化的影响。

参考文献

[1]张鹏举，薛剑，范桂芳. 空间引导功能[J]. 建筑学报，2010（4）：80-84

注：此文摘引自《新建筑》2015.05

Gao Xu From Skin to Expression

Abstract: The extension of the Architectural Hall in the Inner Mongolia University of Technology was completed under a strong character-oriented circumstance. The architect dealt with the issue of emotional space expression through rational methods. By eliminating the original skin materials and manifestation modes from the old building, the common character was generated between the new and the old by the construction method "masonry". And by selecting concrete masonry structure, thinking in honest logic of construction, using a series of separation, transition, opening, and the smooth flow of traffic, it established mutual dialogue and benefits between the new and the old in stable and honest expressions.

Keywords: The Architectural Hall, Extension, Masonry, Skin, Expression

I. The Analysis of the Background

The core of expanding and continuous construction of the architecture, whether concerns with the function or the form, lies on relationship. After being used for four years, the Architectural Hall of Inner Mongolia University of Technology gained the opportunity of expanding for the purpose of containing the design rooms of four undergraduate majors, which would be convenient for teaching and management. In terms of this program, the function is certain, the site is certain, but it is uncertain that what relationship will be between the new architecture and the old architecture for the emerging of the new architecture in environment formed by the old architecture? And how the space appearance of the new architecture will be exhibited? These will be the key priority of expanding the new architecture, and also will be the difficulties.

The expanding of the Architectural Hall of Inner Mongolia University of Technology was continued by the Professor Zhang Pengju, architect of the old architecture. The old architecture was reconstructed by a group of old factory plants in the university in 2009. As soon as it was completed, the old architecture was an "elder" with ages. Its space clearly carried on the historical memory of old industry architectures, which was also an important issue of reconstruction of the old architecture. However, in the design of expanding of the new architecture, a kind of material deciding the appearance of the old architecture-the red brick-is restricted by discipline, which is equaled to closing a direct and effective door on strategy. The architect could only choose the so called "contrastive" method, which is also known as the "modern" technique door. However, this door is widely opened, which like using the character of light and graceful to compare with the character of calm and steady, or using simple to compare with complex; or using the construction of steel to compare with the construction of brick in terms of material, etc, and at the same time, the compare domain is very extensive.

II. The Analysis of the Strategy

In fact, this contrast has already existed in the dimension of campus, so there is no need for struggling. What's more, the light steel structure is not the best choice for teaching building with limited cost. Therefore, the architect makes the identity of the character of space as the emphasis of design. After closing the technique door of continuing to use the same material, this identity naturally resorts to the expression of space emotion.

New space is based on the new function. If red bricks were used, it would only perhaps for the visual strategy of surface, and the expression is presentation of the character under surface. A clear truth is that different surfaces can have same expressions. In other words, walking in spaces with the same expression, the people who experience will naturally ignore change of surface material.

Then, what expression will be continued in building new architecture? The answers given by the architecture are: heavy and real. Heavy comes from both the material and the sense of history of old architecture. Real mainly comes from the direct corresponding of its construction and surface material. Therefore, these two characters both point to the same construction technique-masonry. The construction of the new architecture chooses concrete precast blocks and walls. It is both the construction material and the ultimate surface veneer material.

In response to space logic of masonry structure, the plane layout adopts the standard units from bottom to top. Among six units with distribution of the mean value, five are function classrooms, one is auxiliary space. There is enough clearance between each unit. Among four clearance spaces of north-south confrontation, the two in the north were arranged as stairs and the two in the south open to outside where old trees can connect with sight, adding the relax and communication space and breaking the dull of aisle space. This is the important design strategy of old architecture, and the significant previous for forming the same space character as well. On the first floor was function unit connecting parts of the old architecture where the exhibiting space took shape after unstructured walls were cleaned. Viewing exhibition at the walking line where passengers must pass by also continues one another technique of the old architecture.

The inside space of old architecture is rich and open. Due to function requirement and limitation of construction system, inside space of the new architecture seems a bit dull. Thus, the architect gives priority to the connecting part of the old architecture and the new architecture, and pays attention to outside space. A little courtyard was built in each of the two sides of the aisle connecting the old architecture and the new architecture: the western courtyard seems a bit closed and becomes the quietest one of all courtyards; the eastern one appropriately opens and blends in the campus. Showing the same origin in nature as several ones of the old architecture, these courtyards together become extension of the communication space of professional teaching. Such is the function, and it is especially true with the character.

At the same time, the appropriate separation makes a transition of great disparity in size between old architecture and new architecture and becomes the meditation of scale transformation. In order to intensify effectiveness of this kind of transformation, the new architecture adds an entrance in northwest and extends outward a building built of red bricks. The building and the old architecture create a smaller courtyard suit to original scale. Adding the sense of mystery, the running water is set in the courtyard which is the same as the old architecture. The above techniques are accord with and continue the character of old architecture.

Any strategies to be adopted by the continuous construction of new architecture should not be existed alone. In terms of choice of subject material and structure, along with the consideration of the cost of construction, the masonry structure chose by architect cost less than the reinforced concrete structure, which coincide with the important character of low cost of the old architecture.

Although experience of character is a process from visual to mind, the whole feeling of several kinds of sense organs count as well, like fresh and cool brought by air flow helps intensify the transformation process from eyes to mind. Especially after experiencing ventilation facilitated by passive technique in the old architecture, its function should not be ignored. Imagine if this kind of fresh and cool was not be continued in the new architecture, all kinds of feelings would not be felt. Therefore, in the design of the new architecture, besides convey of space expression, beneficial attempts were also made on ecological strategy for passive space ventilation. The clearance and space between each unit in plane is not only the place for information communication and linking up internal and external areas, at the same time, but also the important place for ventilation. Under the premise that aisle space based on efficiency block draughts, the architect design the vertical ventilation, which uses clearance and space between each unit, provides distribution of airflow by heat pressing, sets air flow path in each layer alone and avoids neutralizing reverse of airflow under the space.

III. The Conclusion

Above all, the continuous construction of the new Architectural Hall carries on the character of moderate, simple and unadorned of the old architecture. The honest expression conveyed by the interaction of space, material and construction of the new architecture make the new architecture and the old one contrast and complement each other. As the professional teaching venue for architecture, the architectural elements of the new architecture, such as space, environment, place, structure, material and etc, would have subtle and formative influence on the students inside.

Bibliography

[1]Zhang Pengju, Xue Jian, Fan Guifang. Space Directs Functions [J]. Architectural Journal, 2010(4):80-84.

Note: It quoted from *New Architecture* in May, 2015.

白丽燕 空间的回溯与期待
——记内蒙古工业大学建筑馆

摘要：通过对内蒙古工业大学建筑馆设计之初的愿景回溯与如今使用现状的问卷调查，印证其对不同使用者的空间意义，并提出在类似开放多义空间的使用中，多层次领域感的确立，是支持各级交往行为发生的充要前提。

关键词：多义空间;建筑馆;使用;期待

一、回溯愿景——从校园活动中心到建筑馆

在内蒙古工业大学校园的中部有一座建于1960年代末的废旧厂房。这组厂房是依据当时生产线建造的铸造车间，到1995年，整体处于废弃状态。2008年初，学校决定改造这组车间，我们对改组旧车间最初的定位建议是校园文化活动中心，在随后的历史机缘和功能探讨中，发现这里是一个天然的建筑馆：铸工车间通透开敞的大空间、自然裸露的结构构件、不加掩饰的结构细部都能够与建筑学重交流、重体验、重实践的教学特点相适应。这一功能也部分地实现了具有开放属性的校园文化中心的原始初衷。2012年，内蒙古工业大学建筑馆在使用4年之后获得了扩建机会，新馆延续了老馆的空间表情，补足了老馆所欠缺的设计教室。

建筑馆这样立足于交流的开放空间中，动线中的节点成为关注的目标，它们是交流的场所，是故事的发生地，同时也是另一类"教"与"学"的场地。在此，线中有点、点中有线，甚至线即是点、点即是线。院子是点、入口是点、中厅是点、楼梯是点，当线中有了"桥"，"桥"也变成了点。

二、使用现状——基于扎根理论的问卷调查

为更加客观地探寻建筑馆的存在对使用者的意义，在扎根理论指导下做了一次面向全校学生的问卷调查。首先将调研对象分为建筑学院的学生与非建筑学院的学生两类，然后将使用者对建筑馆的关注点分为感知印象、空间使用频率、空间喜爱度、物理环境评价四个部分，将两种类型的调研人群对每个关注点的选项频率进行分析对比，并对调查人表述的关键词进行提取、编码与转译，最终得出结果。

1.感知印象
不同学院的学生对建筑馆拥有几乎相似的感知印象，从折线图中可以看出两组折线的走势几乎相同。唯一不同之处在于，建筑学院的学生对建筑光影变换有一定的专业敏感性，因此对该选项两类人群呈现较大的差距。

2.空间使用频率
不同学院的学生在对建筑馆的空间使用频率的折线走势上相差较大。因课业原因，教室是建筑学院的学生使用频率最高的空间，在调研中发现，A馆的休息、评图与展览空间之类的多义空间，并不像预想中的被高频率使用。对非建筑学院学生而言，使用频率最高的则是艺术沙龙，超出我们预料的是，沙龙的使用频率甚至高于本学院的学生。使用频率其次的是A馆的休息与展览空间，同样因为空间的开放与多义，A馆空间对其他学院学生的吸引力使其在某种程度上承担着校园公共社区的功能。

3.空间喜爱度
就喜爱度而言，沙龙与A馆休息空间以绝对优势得到全校学生的喜爱。尤其对于建筑学院的学生，上述沙龙的使用频率虽然不高，但学生对沙龙的喜爱度却很高。入口玻璃厅、展览空间等多义空间也得到全校学生不同程度的喜爱。

4.物理环境
对于建筑馆的物理环境的评价，建筑学院的学生对采光、通风以及室内热舒适度都表示满意，但部分同学对室内声环境不满意，A馆上下通透的开放视野与淡化分层的设计虽然带来了丰富的交流空间，但也必然会导致隔声效果的不理想。非建筑学院的学生对物理环境的选择相对保守，可以看出其选项频率的折线走势较之本学院的平缓很多，仅对隔声问题与本学院学生有不同意见。在对待隔声问题上，建筑学院的学生在建筑馆的行为以学习与交流为主，学习行为对声环境有较为严格的要求，而交流则不可避免地带来各种声音，这是一对带有矛盾属

性的行为。当两者同时发生时，建筑学院的学生必然会对建筑馆的隔声效果持更敏感的态度。而其他学院的学生多以客人的姿态参观建筑馆，其行为以娱乐活动为主，对待声环境的要求亦相对宽松。

5.调研结论——出乎意料的结果
根据对不同学院学生的问卷调研，我们提炼关键词并分析可得学生目前对建筑馆的评价与建议，筛选得出与设计初衷不同之处，令人关注的调研结果。

（1）管理导致的空间使用局限
建筑馆设计有非常充足的出入口，且每个出入口均对应一片精心设计的室外活动场地，其干净整洁的室内外环境足以承重滑旱冰、打羽毛球、开演唱会等各种形式的校园活动。但在此后的管理中，建筑馆日常开启的入口十分有限，此种管理模式一定程度上抑制了上述活动发生的频率与可能性。

（2）空间开放性与私密性的矛盾
学生对建筑馆的空间评价各不相同甚至相互矛盾，比如，数据显示，对艺术沙龙的使用频率外院学生是大于本院学生的，但从空间喜爱度来看，本院学生对沙龙的喜爱度却很高。进一步调查发现，很多建筑学院的学生会有"我以建筑馆为荣，但我对它的使用并不多"的现象，这显然与设计初衷是相违背的。

（3）多样化的学习行为需要多层次的空间
调研中，学生对学习空间提出的需求多种多样，有需要安静独立的学习空间，亦有需要开放性的交流空间。学生表示平时的学习有自习、上课、评图或观看展览等多种方式，他们会根据学习方式的不同而适应性地选择不同空间。同时，学生也希望在A馆的多义空间中能有更多具有停留性与领域感的私密空间，而在B馆的专业教室，能有更多具有归属感的空间存在。

建筑馆由原本杂乱的废旧工厂转型为今天可以承担丰富校园活动、悦纳各种人群的建筑，这其中有我们美好的设计愿景。通过调研发现，一部分空间实现了我们最初的愿望，但仍有一部分空间因为不同的原因而未能发挥其作用，其空间的潜质处于隐藏状态。

三、空间期待——建筑馆的使用潜质

回溯当初设计建筑馆时的初衷，再对比如今建筑馆已发生的空间行为，此次的问卷调查使我们对建筑馆的空间拥有超越一般教育意义之外的理解，这其中既有期待，亦有惊喜。

1.体验与认知
建筑馆建成后，成为校园中最具光阴体验感的场所，饱满的场景与朴素的材料是学生摄影时偏爱的取景场地，也是毕业时拍毕业照的必选之地。同时，作为教育空间的建筑馆也为建筑教学带来了新的视野。上述调研显示，不同学院的学生对建筑馆的光影、空间丰富度均有充分的体验感。但与其他学院的学生相比，建筑学院的学生对空间有更具专业化导向的理解，甚至由于长期浸淫在丰富空间中，学生产生了令人意外的空间分析能力。多年来，建筑馆丰富的空间和质感也激发了老师、同学的创作热情，学院因此开设空间感知与创意素描课程，建筑馆已经成为同学们写生和创作的源泉。在这里，初学艺术的学生可以画它的天光物态、空间结构，练就过硬的写生能力；开始寻求创作的学生运用建筑馆内的任意空间和细节元素解构、重构，画出一幅幅创意十足的作品，建筑馆就是他们艺术再创作的对象。在建筑馆所营造的空间中，学生将入目所及之画进行创意式的重构，为空间感知与素描课程提供新的思路。

2.人文社区
作为校园特色的建筑馆不是建筑学院学生的专属空间，设计初衷是使建筑馆同时承担着校园社区中心的功能。调研发现，由于管理问题，建筑馆的室外活动场地没有被完全利用起来。期待日后建筑馆的管理制度进一步完善，室内外活动场所能被充分地使用，以其开放性的空间、趣味性的功能与独特的艺术氛围引发来自全校师生的交流。对于建筑学院学生而言建筑馆将作为一张自我介绍的名片，而对其他学院的学生，建筑馆则是一张热情的邀请函，一座毕业后的灯塔，时间即使流逝，建筑馆依然具有点燃凝聚力的能量。

3.开放空间中的领域性
建筑馆面向丰富的使用群体，企图满足不同群体对交流空间的使用需求，对空间的尺度把控十分重要。私密与开放是交流生成的空间诉求，因此提倡以合理的尺度和软处理营造出开放的领域性,从而引导丰富的交流行为。比如调研中我们发现,B馆每一层的裱图室中存在着意外丰富的交流性,其提供的领域性空间使学生在裱图等待的过程中有充分交流的可能。此外,建筑学院学生对A馆艺术沙龙的使用率低,但喜爱度却很高的差距也引起关注,分析原因认为,艺术沙龙的存在意义似乎大于其对建筑学院学生的使用意义,即使学生不去使用沙龙这个空间,但路过旧馆时能看到沙龙温暖的灯光,闻到爆米花甜腻的味道,在这个程度上艺术沙龙已经与学生发生了交流,这种另类的交流形式值得进一步的研究。

追求空间的开放与交流是设计的一个方面,但作为教育建筑,建筑馆必然要有适合学生学习的空间。调研显示学生对学习空间有着看似矛盾的空间需求是因为学习行为的多样化。比如,需要相对独立空间进行的自省型学习;需要开放空间的交流型学习;需要教学空间的教育型学习。因此,学生既需要开放性的评图空间,也需要24小时独立教室。建筑馆需以多层次的学习空间来满足学生多样化的学习行为,开放空间也需要确定不同层次的领域感和舒适度来适应不同的使用人群。

结语

内蒙古工业大学建筑馆自建成使用至今已有九年时间,作为内蒙古工业大学建筑馆成长的见证人,感受到好的建筑本身会具有灵魂,具有感召力,让身处其中的人不自觉地受到其影响。正如查理·芒格所言,"如果你想得到什么,最好的办法是让你自己配得上它"。九年以来,内蒙古工业大学建筑学院逐渐积累与进步,并获得很多"意外"的机遇。通过这次对建筑馆成长的回顾和思考,似乎是我们与工大建筑馆共同成长中配得的奖励。

参考文献

[1]张鹏举.适应·更新·生长——一次人文与生态事业下的旧产业建筑改造实践[M].北京:中国建筑工业出版社,2011.

注:此文摘引自《世界建筑》2017.07

Bai Liyan Retrospect and Expectation of Space
—— Record of Architectural Hall of the Inner Mongolia University of Technology

Abstract: Through the questionnaire survey of the vision of the architectural design of the Architectural Hall of the Inner Mongolia University of Technology and the retrospect of the present situation, it confirms its spatial significance to different users and puts forward the establishment of multi-level domain sense in the use of similar open polysemy space, and it is to support all levels of communication occurred in the necessary prerequisite.

Keywords: Polysemy Space, The Architectural Hall of the Inner Mongolia University of Technology, Use, Expect

I. Looking Back upon the Past Wish— from the Campus Culture Center to the Architectural Hall

There is a deserted factory, built in the late 1960s, in the middle of the campus of the Inner Mongolia University of Technology. This group of buildings was the workshops according to the previous production lines at that time, and the whole was deserted by 1995. At the beginning of 2008, the university decided to reorganize these workshops, and the initial usage of reorganizing workshops was confined as the campus culture center. With the following opportunities and function-discussion, we found out that it would be a natural architectural building: the spacious place, exposed components and uncovered details of the workshop can be best suitable to the characteristics of teaching architectural engineering that is, focusing on interaction, experience and practice. And this function partially fulfilled the original purpose of opening characteristic for the campus culture center. In 2012, the Hall has obtained the opportunity to extend after being used for four years, and the new building has continued the space expression as usual and made up with the newly-built designing classrooms. It is far more enough for the Architectural Hall aiming at interaction. Furthermore, the circulation joints become the focuses; they are the places for communication, the places for stories, meanwhile, and another kind of teaching and studying. Here, the lines and spots coexist with each other, even can substitute each other. The spots are yards, entrance, hall and stairs, when the bridges are in the line, they become the spot too.

II. The Current Using Situation — a Questionnaire Survey based on the Grounded Theory

In order to probe objectively the meaning and importance of the users towards the Architectural Hall, guided by the Grounded Theory, the authors conducted a questionnaire survey to the whole school students. On one hand, the survey subject is divided into two parts: the students from the College of Architecture and the students from the other colleges; on the other hand, the focusing points the users towards the Architectural Hall fall into the following four categories: the perception impression, space-using frequency, spatial preference and the physical environmental assessment. The final results can be drawn by comparing, analyzing every focusing point frequency by the subject, in the meantime, extracting, encoding and translating the keywords of the interviewees.

1. The Perception Impression

There is seemingly similar perception impression of various college students towards the Architectural Hall, and it can be seen from the broken line graph that the two groups of broken lines run more or less the same. The only difference lies in that the students from the College of Architecture have certain professional sensitivity of the changes in the iridescence of architectural lighting, and therefore, this item varies greatly between the two kinds of subject.

2. The Space-Using Frequency
There is a great deal of difference, according to the tendency of the broken lines in the space-using frequency of the Architectural Hall, among different college students. Due to the reason of the need of courses, the classrooms are the most-used place for the students from the College of Architecture. The survey shows that the polysemy space like the resting, design review and the exhibition place of the A-Building is not highly used as we imagine. As far as the students who are not from the College of Architecture is concerned, the Art Salon is the place they visit and use most frequently, and far beyond our expectation, the frequency is even higher than that of the Architecture students. The using frequency of the resting place and the exhibition place in A-Building follow the Salon and the three places have the common reason is that the space is open and polysemy. To some extent, the attraction of the A-Building towards the other colleges' students makes it serve the function of the public community on campus.

3. The Spatial Preference
In terms of preference, the salon and the resting place in A-Building, which have absolute advantages, won the preference and favor of all students. Especially for the students from the College of Architecture, the Art Salon is the place they have huge likability on, even though they visit and use not so frequently. Also, the polysemy space like the glass hall nearby the entrance and the exhibition place draw various degrees of all the students' attention and preference.

4. The Physical Environmental Assessment
As for the physical environmental assessment of the Architectural Hall, the students from the College of Architecture are very satisfied with the lighting, ventilation and the indoor thermal comfort, whereas some are not so pleased with the indoor sound-system. Even though the transparent open vision and layer-weakening design of A-Building brings rich interaction space, however, it inevitably leads to the undesirability of sound insulation effect. However, the students from other colleges are more conservative in physical environmental assessment and we can see that their tendency of the broken lines is more steady and smooth, with only difference in sound insulation compared with the students from the College of Architecture. Dealing with the problems of sound insulation, the students from the College of Architecture hold the opinion that their behaviors include study and communication. Study has very critical requirements for environment, whereas communication inevitably can cause different kinds of noise, and they are a pair of contradictory behaviors. With the moment they take place simultaneously, the Architecture students are bound to have a negative and sensitive attitude towards sound insulation effect. While the other students visit the Hall as guests and their behavior is mainly about entertainment, so relatively speaking, they have mild and low requirements for sound environment.

5. Research Conclusion—Unexpected result
According to the questionnaires for different colleges' students, we chose the keywords and analyzed their current assessment and suggestion towards the Architectural Hall. Then we highlight the ones which differ from the original purpose and draw the conclusions which are of great concern.

(1) Limitation of Space Use Caused by Administration Management
The Architectural Hall has plenty of entrances and exits which are all connected with well-designed outdoor activity space. Their clean and neat indoor/outdoor environment is suitable enough to hold all kinds of campus activities, such as roller-skating, playing badminton, opening concerts, and so on. However, during the process of administration management afterwards, the number of the opening entrances to the Hall is very limited and such management mode, to some degree, prohibits the frequencies and possibilities from the above activities.

(2) The Contradiction of Space Openness and Privacy
The students' assessment towards the space of the Architectural Hall differs from each other, and some even contradictory. For example, the data shows that the usage frequency of the Art Salon for other colleges' students is higher than that for Architecture students, but from the perspective of the spatial preference, the students from the College of Architecture show much more interest. Further investigation finds out that most Architecture students hold the idea that "I am proud of the Architectural Hall, but I don't use it too much" which is obvious contradictory to the original intention of the design.

(3) Various Learning Behaviors Need Multi-Level Space
In the process of survey, students raised a variety of requirements for learning space, some hoped quiet independent learning space and some hoped open communicative space. Students said that the usual study included the following ways: self-study, taking-lessons, design review, watching the exhibition and so on, and they would choose the corresponding space. At the same time, they hoped to have more privacy space with the characteristic of plausibility and the sense of territory in the

polysemy space of A-Building, whereas more belonging sense of space existence in professional classrooms of B-Building. The Architectural Hall, transformed from an old disordered and deserted factory to a new campus activity center capable to accept and welcome different kinds of people, has immersed our promising and prosperous design vision. Through the research, we find that some space fulfilled the original purpose, while for different reasons, the other did not play a powerful role and the potential still was in the state of invisibility.

III. Expectation of Space —— the Usage Potential of the Architectural Hall

In terms of the Architectural Hall, looking back upon its original purposes and compared with its present space behavior, we have an expecting and surprising understanding of its space which is far more beyond normal education meaning by means of the questionnaire and survey.

1. Experience and Perception

After being built, the Architectural Hall has become the most favorite and attractive place to appreciate the time. With its plentiful scenes and simple material, the hall draws students' attention and favor to go in for photographing; also it is a must of graduation photos. At the same time, as an education place, the hall has brought some new vision and insight for architecture teaching as well. It can be drawn from the research that all of the different colleges' students have full experience of the light and spatial richness of the Architectural Hall. But compared with other students, the students from the College of Architecture have more professional sensitivity and understanding. Even because they have been staying in rich space for a long time, students possesses the unexpected spatial analysis capability. For the past many years, the rich space and sense of reality of the hall has aroused much creative enthusiasm for faculty and students, thus the college has offered the Space Perception course, the Creative Sketch course and the hall has become the birthplace for students' painting and creation. Here, the beginners can draw its items, details, structures and achieve their excellent sketching ability; the ones who purse creation can use any space and detail elements for deconstruction and reconstruction, because the hall is just the portrait and model for their artistic re-creation. With the space created by the hall, students reconstruct creatively everything they see and provide the new thought for offering the Space Perception course and the Creative Sketch course.

2. Humane Community

The Architectural Hall, unique feature of campus, is not the private space for Architecture students. The original purpose is to make it fulfill the function of the campus community center as well. It can be seen from the research that the outdoor activity sites of the hall haven't used fully due to the administrative management. We are looking forward to see that the managerial system will improve step by step, the indoor and outdoor activity sites are going to be used fully and communication from all the faculty and students is to achieve with its open space, interesting function and unique artistic atmosphere. As far as the students from the College of Architecture is concerned, the Architectural Hall is just like their own name cards; while for other colleges' students, the hall is like a piece of warm invitation letter and a beacon after graduation which has and will have the power to unite with time passing by.

3. Territoriality of Open Space

In the face of a variety of target groups, the Architectural Hall tries to meet their different needs for space communication and it is very important to control the scale of the space reasonably and satisfactorily. Privacy and openness is the space pursuit root for communication, therefore, we call for sparing the territoriality of open space with reasonable scale and harsh treatment so as to make various and colorful communication behaviors. For example, we find out during the research that the framed drawing room in every floor of B-Building is filled with extremely meaningful and insightful communication, and its territoriality of open space makes it possible for students to communicate fully while they are waiting for their framed drawing. Besides, the thing which also draws attention is that the architecture students have a low using-frequency but a high favor of the Art Salon in A-Building. We analyzed the reasons and draw the conclusions: the existing meaning of the salon for architecture students is far more significant than the using meaning for other students, and the architecture students can see the warm light and smell the sweet popcorn passing by even though they do not use it. In this sense, salon has already communicated with the students and such unusual communication way needs further research.

Seeking openness and communication is one aspect when we design the space; however, being education architecture, the Architectural Hall must have plenty of space which is suitable for students' learning. The seemingly contradictory needs students have towards the learning space shown in the research comes from the variety of learning behaviors. For instance, introspective learning needs relatively independent space, communicative learning needs open one and educational learning needs teaching space. Therefore, not only do students need the open space reviewing designs, also they require the independent classroom for twenty-four hours. To sum up, the Architectural Hall is supposed to satisfy the students' various learning ways by offering multi-level learning space, and the open space should also be worked out for different target groups by different levels of territoriality and comfort.

Conclusion

It is nine years sine the Architectural Hall of the Inner Mongolia University of Technology has been built, and being the witnesses of its' growth, we really feel that the good building and hall like the Architectural Hall has its own soul, charm and charisma which influences the visitors and users subconsciously. Just like what Charles Thomas Munger said: "Safest way to get what you want is to deserve what you want." Also for the past nine years, the College of Architecture in the Inner Mongolia University of Technology has accumulated and improved gradually, and achieved many unexpectedly surprising opportunities. In the process of this recalling and thinking, we feel honored and rewarding that we grow up along with the Architectural Hall.

Bibliography:
[1]Zhang Pengju. Adjustment· Improvement· Enlargement—An Old Factory Renovation in Views of Humanity and Ecology [M]. Beijing: China Architecture & Building Press, 2011.

Note: It quoted from *World Architecture* in July, 2017.

四、建筑馆的画和话
IV. The Pictures and Utterances of the Architectural Hall

建筑馆的画
The Pictures of the Architectural Hall

学生：庞博文　　指导教师：刘洁
Student: Pang Bowen　　Supervisor: Liu Jie

建筑馆写生

清晨，建筑馆的天光教室迎进天空的第一抹晨光，整个馆内被映得红彤彤，透着一股子朝气。沿着楼梯拾阶而上，惊起一只鸟儿在梁间扑腾翅膀，嗖的一声不见了踪影。抬头间，玻璃窗形成的十字格墙、笔直的角钢与三角形水泥提梁相互穿插交错，构成结构丰富的空间。灰色高高的天顶被时光晕染成千变万化的水墨画，痴呆呆地一直望着，便看到了精彩的抽象作品。

一天的时光，光影流转，阳光透过各类窗户停留在红砖墙上。时光一分一秒地过去，一道道大梁的影子由浅至深，顺着砖缝倏然不见了。东边工字钢交错的穹顶在夕照下更显得结构繁密，一不小心美丽花边般的投影在黑色钢炉的管壁上，坚硬的黑冷泛起温柔的光。至晚，银色的圆灯一盏盏亮起了，建筑馆在夜色的衬托下显得格外肃穆。

最喜欢春日里从三楼画室的窗前看建筑馆前后的树吐出绿芽，粉嫩的绿更衬得砖红色的建筑生机勃勃。夏日里窗外炎热，花开得热闹，一走进玻璃厅一股凉爽就扑面而来。长长的白日，光影的变化是建筑馆墙壁上一层生动的墙饰。最美是秋日的建筑馆，"爬山虎"爬满一墙红砖，院子中央雕塑般的大烟囱矗立在一片深深浅浅的橘色中，从三楼画室向外望去，满地的落叶，满桥的枝蔓，童话般的色彩和韵律。冬日里萧索的校园，一走进来，最夺人眼的就是红砖砌就的建筑馆，东边高凸的屋顶钢构件泛了红锈，活脱脱一双向外望的眼……

红色粗粝的砖、灰色温柔的水泥、黑色坚硬的金属、浅灰光滑的地面泛着光，甚至一枚扎进墙里的钉子，一段平行直率的水管都构成了眼前这个质感世界。在这个建筑中学习生活，是环境影响了人，抑或是人造就了环境，渐渐地，里面的人有气质了，这建筑也活了起来。

多年来，建筑馆丰富的空间和质感激发了老师同学的创作热情，建筑馆已经成为同学们写生和创作的源泉。建筑馆内处处是景，在这里初学艺术的学生可以画它的天光物态、空间结构，练就过硬的写生能力。在这里开始寻求创作的学生运用建筑馆内的任意空间和细节元素解构、重构，画出一幅幅创意十足的作品，建筑馆就是他们艺术再创作的对象。在这里师生们感受到自然与建筑的和谐共生，真可谓：目之所及皆可入画，在艺术的氛围中，成长艺术。

刘洁

Sketched from Life of the Architectural Hall

In the morning, the first light came into the Sky Classroom of the Architectural Hall, making it filled with bright red lights and vitality. A bird in the beam was startled by the sound of going up stairs. It fluttered wings and flew away suddenly. Cross-grid walls of glass windows, straight angel steels and triangular cement beams were crisscrossed, constituting a space with rich structures. Lights were moving, painting the grey high ceiling as an ever-changing picture. If staring at this, an abstract painting appeared gradually.

One day, lights flowed, stayed on the red brick walls through various types of windows. With time past, shadows of beams were from shallow to deep, disappearing along the brickworks. The crossed I-shaped steel dome on the east seemed denser structured in the sunset, projecting lace-like borders on the pipes of black steel furnace, softening the tough and dark equipment. To the evening, silver round lights lit. The Architectural Hall was more solemn in the night.

It's my favorite thing to see trees around the Architectural Hall to burgeon forth in spring through the studio window on the third floor. The fresh green sets off the red-brick building full of vitality. It's quite hot in summer, and flowers are blooming lively, but you will feel cool as soon as you walk into the glass hall of the Architectural Hall. From sunrise to sunset, light changes are vivid decorations of building walls. The Architectural Hall has the most beautiful scene in autumn. The red brick walls are covered by ivy, the big chimney stands in shades of orange in the center of the courtyard just like a sculpture. There are full of fairy tale color and rhythm when you see the fallen leaves and branches on the ground and the bridge through the studio window on the third floor. The campus is desolate in winter; however, the red-brick Architectural Hall is still the most attractive when you walk into the campus. The high convex steel members of the east roof have red rust, which look like a pair of eyes that are looking out at the views.

The rough red brick, soft grey cement, hard black metal, and smooth light grey ground suffuse with light. Even a nail into the wall or a parallel straight pipe is an important part of the texture of the world in front of you. It is hard to tell whether people create the environment or the environment has influence on people when learning and living in this building. Gradually, people in it have the temperament, and this building is also alive.

Over the years, the rich space and texture of the Architectural Hall has stimulated the creating enthusiasm of teachers and students. The building has become a source of painting and writing for students. Because there are amazing scenes everywhere in the Architectural Hall, fresh art students can draw the skylight state and its spatial structure here so that they could acquire excellent sketching ability. Students who seek chances of creation come here to make use of the building space and to deconstruct and reconstruct the details of the construction elements, and then they paint lots of creative works. The Architectural Hall is the object of their art recreation. Teachers and students can feel the harmony of nature and architecture here. It is that you can draw everything you see, and the art is growing in the artistic atmosphere.

Liu Jie

学生：贾伟　　指导教师：刘洁
Student: Jia Wei　　Supervisor: Liu Jie

学生：陈晓旭
指导教师：刘洁
Student: Chen Xiaoxu
Supervisor: Liu Jie

学生：宋倩
指导教师：刘洁
Student: Qing Qian
Supervisor: Liu Jie

学生：凌梓瑄
指导教师：刘洁
Student: Ling Zixuan
Supervisor: Liu Jie

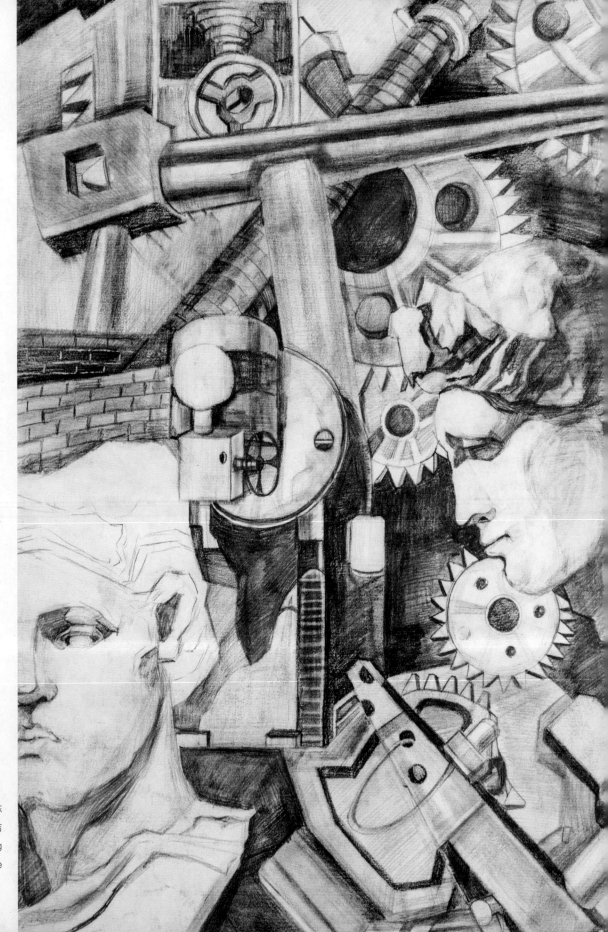

学生：齐栋
指导教师：刘洁
Student: Qi Dong
Supervisor: Liu Jie

学生：王雅娜（上）
指导教师：刘洁
Student: Wang Yana
Supervisor: Liu Jie
(Above)

学生：胡晓云（左下）
指导教师：刘洁
Student: Hu Xiaoyun
Supervisor: Liu Jie
(Lower Left)

学生：齐栋（右下）
指导教师：刘洁
Student: Qi Dong
Supervisor: Liu Jie
(Lower Right)

建筑馆的话
The Utterances of the Architectural Hall

学生：伊菲菲　指导教师：刘洁
Student: Yi Feifei　Supervisor: Liu Jie

学生：王笑颜　指导教师：刘洁
Student: Wang Xiaoyan　Supervisor: Liu Jie

崔愷

早几年应邀去内蒙古工业大学院建筑学院讲座，在刚刚完成的老厂房改造而成的建筑馆中体验。质朴的材料，精准的细部，丰富的空间，精心保护和巧妙利用的工业遗存都散发出强烈的力量感和震撼力，记得我当时向同学们有感而发：这是我见过国内最好的建筑学院的大楼，是出好建筑师的地方！后来我还带着院里建筑师们去张鹏举先生在乌海的青少年创意产业园学习交流，那也是个老工厂改造，朴实的红砖从墙上一直铺满院子，高高低低，错错落落，一丝不苟，让建筑与大地长在了一起，也与四面的沙山十分默契，很大气，很平实，同去的建筑师们都有同感。

赵辰

对于工业遗产，相对于保护，我觉得更重要的问题可能是利用，即：要研究、规划、设计工业建筑空间的重新再利用问题。对此，内蒙古工业大学建筑馆是最充分的、最有力的实践证明。

校园建筑本身是最典型的，最具教化意义的建筑，这种建筑能够代表一种新的发展理念，它与校园的重新融合恰恰是对学生最好的教化。对于内蒙古工业大学建筑馆，在这里受教育的学生，并不只是在课堂里可以学到知识，他也可以从实物里面体验到一些东西，而建筑学的很多知识正是需要这样学的，我觉得非常好，所以，我觉得这才是真正需要保护和复兴的内容。

以上这些想法与体会是我对内蒙古工业大学建筑馆精彩案例的回应。

孙一民

呼和浩特，古城绥远，静静伏倚在阴山下，远离国内的仍频大事件，甚至没有迎合内蒙古近年冒进般快速发展，内蒙古工业大学建筑学院就在这红砖厂房的院落里，凝练着自己的理想与精神。

只有当你走进那外表普通的半敞开院落，进入厂房内那精心安排的教学空间，看着工业建筑粗大构件的阴影将明亮的阳光洒向设计课的每一个课堂，体会同学们在温暖宜人的空间下思考设计，你会意想不到地被感染。在这里，专业、年级、老师、学生之间的隔阂消失了，教与学的关系，回到了大树下围坐切磋的古希腊理想氛围之中。

建筑塑造的精神，好的建筑会引导人纯净，这几乎成了建筑师无法实现的理想。但看着这个在安静中潜伏着厚重的建筑学院，我相信，这里的同学将会在未来，在走出院久以后，体会到质朴、坦诚的建筑精神已深深烙印在心灵的深处。

张颀

在初建车间那个年代，工业建筑的特征表达得非常清晰，外观非常质朴，主要是适应了当时的工业生产。改造后的建筑馆，无论是从气度上，还是装饰上，乃至技术上都将这些特征延续了下来，包括墙的建造方式，也采用了传统的一顺一丁的砌筑方式。整个项目做得非常优秀。我认为这个改造项目还达到了两个目的：首先是关于交流，该项改造实践创造了丰富的空间，对学生间、师生间的交流十分有益；其次是关于认同感和场所精神，我认为这个系馆能够在校园里形成一种认同感或者具有某种场所精神。何谓场所精神？简单地说，就是大家在这里工作、学习、生活，会对它形成一种认同。过去，这个车间肯定在这个区域当中是非常重要的，是一个标志，现在，变成这个系馆以后，就成了深受学生喜爱的一个学习环境，而这些学生都是未来的建筑师、艺术家。

学生：陈晓旭　指导教师：刘洁
Student: Chen Xiaoxu　Supervisor: Liu Jie

学生：王雅娜　指导教师：刘洁
Student: Wang Yana　Supervisor: Liu Jie

魏春雨

我接触了很多遗产改造项目，但总觉得表演的成分比较多，包括798。今天来到内蒙古工业大学建筑馆，最大的感受是，里面一点表演的成分也没有，都是实用的，或者言，这些空间都是为了用而做的。这点跟现在外面很多披着遗产保护的外衣，实际上是为了"炫"的东西是不一样的。这种很纯朴的价值观对学生潜移默化的作用会更大，我觉得这点特别好。其实这个挺难做到的。做设计的本能就是要做好看的，这个无可厚非，如果能跳开这个模式去做设计，真的挺让人敬佩的。

另外我想借此谈谈新与旧的问题。我接触的很多类似作品，包括在国外看到的，都特别强调新旧对比。通常是用新的材料来表达骨子里的旧东西，做得特别炫，而且做出的空间极其富有创意，视觉冲击力很强。但是我发现建筑改造项目大量地利用了回收的旧材料，而且这些材料用得很重。这与我们通常用很轻的材料去塑造这样的空间，并利用二者的反差来达到某种戏剧性的做法是不同的。这样的方式使我不能一下子很清晰地识别新与旧，这很耐人寻味。我想，这应该是个价值观的问题。从这个层次上讲，这个改造项目新和旧的处理比一般的我们所看到的披着工业遗产保护外衣的表演要好。

赵崇新

这个改造项目有个最大的优点，就是绿色、省钱。设计者运用了很多原有的东西，这对于我们建筑师而言，是真正需要去研究的，是非常重要的。这个房子改变了我对内蒙古工业大学的看法，甚至是改变了我对内蒙古的看法。我个人认为，一个项目能够改变人对一个地区的看法，这是很不简单的事情。此外，我认为，由于鹏举老师既是设计师又是业主，所以才能够把这个设计做得如此完美。

孔宇航

对于内蒙古工业大学建筑馆，一是废墟美；二是关于光与空间的关系。对于前者，工业建筑在六十年代是非常常见也非常实用的，无论时间怎样变化，品质不变，废墟美是建立在这种品质之上的；关于光与空间，是不应随着改造和功能的变化而变化，都是人永远需要的。这个系馆的设计，光的运用是很优秀的，它打破了一般的建筑师根据规范，或者建筑的常规做法来做建筑的规律。所以，废墟美与由空间与光的关系产生的流动性，这两点很打动我；第三，我觉得对混沌理论最重要的理解就是其不确定性和不可预见性，例如，当初设计工业厂房的时候，大家绝对不可能想到它现在会是建筑馆。这带给我们的思考是什么？就是在建筑学教育中，我们为学生制定任务书时，是否要考虑建筑未来的开放性，或者说功能的不确定性？内蒙古工业大学建筑馆改造项目是很好的例子。即在原有的体系和空间布局下，将其进行不同功能的转换。一个制造产品，一个培养人才，这两种完全不同的功能要求带给我们对教育的反思：在这样的一个时代，我们怎么去考虑设计在未来的不可预测性；最后，是关于形式的反思。建筑师该怎么样去思考形式？这是真正值得我们关注的事情，而不仅仅只是为了带来视觉上的冲击。这个设计是由内而外的设计，我觉得这符合有机建筑的理论。很多建筑师的设计是由外而内的，总是先把形式搞好。我觉得，真正优秀的建筑师一定是由内而外做设计，设计就像生长规律一样，起因于内在的基因。

张应鹏

在这个建筑馆改造项目中，鹏举师兄同时是业主、设计师和监工。在这种三位一体的情况下，他实现了这个房子，我很羡慕他。有这么一个机会，能够把自己的想法发挥出来。我想，这是建筑师人生中最大的理想，也是一种完美的境界。说实话，我认为这个房子是师兄发挥得淋漓尽致的一个。我想，这其实是一个理解和沟通的问题，它导致建筑师的设计能力不能得到充分发挥。现在看到这个房子，我在想，有这么好的一个东西在这里，这么好的一个案例在这里，那么下面的机会肯定会更多。社会的信任度也会随之增加，当然，也包括学校的，讲大一点，呼市的，再宽泛一点，内蒙古的认可度也会提高。

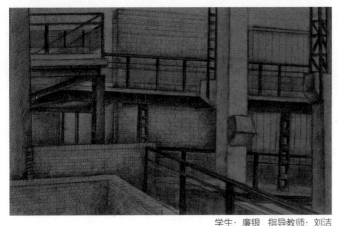

学生：廉银　指导教师：刘洁
Student: Lian Yin　Supervisor: Liu Jie

学生：刘潞　指导教师：刘洁
Student: Liu Lu　Supervisor: Liu Jie

李晓峰

鹏举院长在做这个改造项目的时候，对于旧与新的关系的处理是有自己的想法的。他没有完全把新的东西拿过来掩盖旧的东西，也没有用一些粉饰的东西把它做得很炫、很酷，反倒是将新旧在不同时期的积淀凸显了出来，我觉得这是非常好的。从这个方面来评价这个建筑的话，它既符合我们旧建筑再利用的建筑学改造原则，同时也为未来作为一种文物建筑或者被认定为该类建筑留有余地。

杜春兰

参观建筑馆，一直处于兴奋状态，觉得这才是真正的建筑学院。这里面有许多兴奋点，往上看，是一种感觉，往左看，又是一个风景，这个建筑总是让人有意想不到的惊喜。我想，如果在这个地方还学不出来，那就真学不出来了，所以我很羡慕这里的学生。

王兴田

建筑设计是由内而外的，这应是普遍性规律，内蒙古工业大学建筑馆是一个代表。这个空间是以工业遗产的状态存在，它的功能、空间、形式在六十年代是很常见的，现在，我们需要反过来再去重新认识它在当代的存在。

我认为，这个旧工业厂房内外空间是相当协调一致的，包括空间的比例、大小、尺度以及分割，我觉得设计者将这种感觉发挥到了极致，而这种感觉正是由内而外的具体表现。这个建筑确实让我感动，因为它的另一个特点就是利于师生的交流，在这儿学不好建筑，去哪儿也学不好。另外，我觉得它能够从空间、技术、构造上对学生产生潜移默化的作用。对于这方面，鹏举院长很会动脑筋，他把里边所有留下来的东西都用上，并且用得很好。这说明他们师生在日常生活或学习行为当中都有这样的意识。所以，我觉得这种建筑是很有教育意义的。举个例子来说，这里面的砖结构、木结构以及钢结构，包括这么多结构的这么多节点，对于我们的构造课程是很有帮助的，这要比老师凭空讲述这些内容更容易理解，可以说，这种建筑学的教学方式是最直观的。

Cui Kai

A couple of years ago, I was invited to give a lecture in the couege of Architecture of IMUT and I really had the opportunity to experience the newly-built Architectural Hall reconstructed from the old factory workshop. The plain-style material, precise detail, abundant space, carefully preserved and masterfully exploited industrial heritage releases a sense of empowerment and earthshaking force. I was inspired at that moment and spoke to the students this was the best architecture department building I had seen in China and it must be the good place to create good architects. Then I led some architects in our department to visit and learn the Inner Mongolia Wuhai Youth Creative Industry Center by Zhang pengju which was also a case of old workshop reconstruction. The natural and plain red-bricks covered the whole courtyard from the wall, up and down, intricately and meticulously. And I had the same feeling with those architects that it made the architecture and the field grow together, also there is a tacit agreement between the sand hills all around, bigheartedly and profoundly.

Sun Yimin

Hohhot, ancient Suiyuan, is quietly located under the Yin Mountains. It is usually not affected by big events frequently taking place in China, or even does not cater to the rashly rapid development of Inner Mongolia in recent years. College of Architecture of IMUT is located in a yard in which stands a red brick workshop building, embodying the dreams and spirits of its own.

Stepping into the ordinary semi open yard and entering the well-designed teaching space, you will be touched unexpectedly: the sunshine tenderly filter through the gross component members of the industrial buildings, pouring down its warmth into each design studio where students think about their designs. Here, the gaps disappear between majors, grades, teachers, students, and the relationship between teaching and learning is harmonious, similar to the ancient Greek ideal thinking atmosphere in which Greek thinkers discussed with each other under a big tree.

Architecture shapes the spirit of the building. A good architecture will purify the human's soul, which is the supreme ideal sought by architects all of their lives. Looking at the Architectural Hall with its massiveness and elegance lying hidden in a quiet environment, I believe, for a long time after they have left the college, students here will realize that the simple and honest spirit of architecture has been deeply imprinted in the depths of their souls.

学生：宋倩　指导教师：刘洁
Student: Song Qian　Supervisor: Liu Jie

学生：高雪芳　指导教师：刘洁
Student: Gao Xuefang　Supervisor: Liu Jie

Zhao Chen
Relative to the protection for industrial heritage, I think the more important problem may be to utilize. That is, to study, plan and design rescue of the industrial building space. For this, the Architectural Hall of Inner Mongolia University of Technology is the best and the most powerful demonstration. Campus building itself is the most typical one of the enlightenment meaning of architectures. This kind of building can represent a kind of new development concept. It is merged with the campus, and is the best education for students. For the Architectural Hall of Inner Mongolia University of Technology, students who learn here not only can acquire knowledge in the classroom; they can also experience some things from practice. And much of the knowledge of the architecture is needed to learn in this way. I agree with it, so I think this is the content which we need to protect and renew. These ideas and understandings are my response to the wonderful case of the Architectural Hall.

Wei Chunyu
I have seen many heritage renovation projects, but I always sense that there are too many hints of performance in them, including Beijing 798 Art Zone. Today when I came into the Architectural Hall of Inner Mongolia University of Technology, the most impressive thing I've felt was that there was no any trace of performance in it. Everything was practical, that is, all the spaces were practically useful. That was different from those things under the cloak of heritage protection with the real purpose of "dazzling". In my opinion, it is particularly good that the simple value exerts a big and subtle influence on students, which, in fact is hard to achieve. To design something that is good-looking is out of every designer's instinct, which is understandable. At the same time a designer should be well respected if he/she could get away from the traditional thinking model. In addition, I would like to talk about the issue of new and old. I have seen many similar projects, including some abroad, and all of them had particular emphasis on contrast between the new and old. The essentially old things were usually expressed by new materials, appearing "dazzling", and having a strong visual impact, with extremely creative space. But I found that a lot of recycled old materials were used in this architecture building renovation project, and the materials were quite heavy. This was different from the tradition in which much lighter material was used to shape the spaces since some kind of dramatic effect was achieved by the contrast between the two. For such a way, I couldn't clearly identify the new and old quickly, which was thought-provoking. I think it should be an issue of values. In this sense, what have been done on the old and new in the renovation project is better than some performances under the cloak of industrial heritage protection as we usually see.

Zhang Qi
At that time when the workshop was firstly established, the characteristic of industrial construction was expressed very clearly, and the outlook was very plain. It fit in the industrial production of that time. The reformed Architectural Hall inherits the features technically no matter in bearing and in decoration. The traditional masonry is adopted to build the walls. The whole project is excellent. I think this reformed project achieves two purposes. One is that this project creates rich space for student-student and teacher-student communication; the other is related to sense of identity. I believe this construction can form a sense of identity on campus. Simply speaking, we work, study and live here, and then we will have a sense of identity. In the past, this workshop must have been very important in this area, and it was a sign. Now it turns into a department building, a favorable learning environment for students, who are future architects and artists.

Zhao Chongxin
This transformation project has the biggest advantage that is green and saving money. Designers use a lot of original things, which for our architects, is the real need to study, which is very important. The house changed my view of Inner Mongolia University of Technology, and even changed my view of Inner Mongolia. I personally think that a project can change people's perception of a region, which is not a very simple thing. In addition, I think, because Pengju is both the designer and the owner, and that is why he is able to fulfill this design so perfectly.

学生：吴难　指导教师：刘洁
Student: Wu Nan　　Supervisor: Liu Jie

学生：郭宇轩　指导教师：刘洁
Student: Guo Yuxuan　　Supervisor: Liu Jie

Kong Yuhang

The Architectural Hall of IMUT has many features. First, it is reformed from ruins. Industrial building is very common and useful in the 1960s. No matter how time changes, its quality remains unchanged, which is the base of its ruins' beauty. Second, it utilizes the light and space artfully. These two points will not be altered with its reform or function changes, and will always necessary. The light design of this pavilion breaks the general norms and the architects' usual practice; it is beyond the common construction specifications. So, ruins' beauty and liquidity produced by the light and space are the two points which struck me most. Third, it involves the chaos theory, which implies uncertainty and unpredictability. For example, its original design was an industrial building, so it's hard to imagine it is the Architectural Hall now. Then what can we learn from the change? In architecture education, when students' commitments are delivered, should we consider the future openness of the architecture, or its function's uncertainty? The renovation project of the Architectural Hall of IMUT might be a good answer. Namely under the original system and space layout, the function can be transformed. Manufacturing products and training talents have completely different functional requirements, but they both bring us to reflect on education: in such an era, how do we consider designs' future unpredictability? In the end, the reflection on form is focused. How should an architect think about forms? It is truly worthy of our attention, but not just for the sake of visual impacts. The motive of a design should come from the inside out conforming to the theory of organic architecture. However some architects' designs are from the outside in, and always attach too much emphasis on the form first. In my views, excellent architects would design from the inside out, like growth rule, paying much attention to the inherent genes.

Zhang Yingpeng

In my opinion, I think my dear schoolmate, Zhang Pengju, is the owner, designer and supervisor in this reconstruction project of the Architectural Hall. The trinity achieves him this HOUSE and I admire him a lot because he has such a good opportunity to express his ideas out. And I strongly believe that is the biggest ambition and perfection for every architect. Frankly speaking, I regard this HOUSE as the best one of my dear schoolmate pengju. I consider that this is a question of understanding and communicating, in which makes architects unable to achieve their full potential in designing. Seeing this a wonderful and gorgeous HOUSE, I said to myself that with such an example and model there will be plenty of opportunities and the social credibility must be risen up as well. Of course, including the one comes from the university , to a larger sense, comes from Hohhot even from Inner Mongolia.

Li Xiaofeng

Dean Zhang pengju has his own idea about how to deal with the relationship between the old and the new in this reconstruction project. Neither did he use the new stuff to cover up the old, nor use some decoration to make it shining or cool. As the matter of fact, the accumulation and heritage of the old and the new in different ages was highlighted and I really think it is marvelous. To comment the architecture from this point of view, I assume that it not only meets the principles of architectural transformation by re-using the old buildings, but also leaves the way open for a certain cultural relic build or such in the future.

Du Chunlan

Visiting the Architectural Hall, I have been in a state of excitement and had the feeling that this is the real Institute of Architecture. There are a lot of excitement, look up, is a kind of feeling, to the left, also a landscape. The building always let people have unexpected surprise. I think, if you can not learn in this place, it really does not come out, so I am very envious of your students here.

Wang Xingtian

The architectural design should take an inside-out perspective, and this universal law can be exemplified in the Architectural Hall of the Inner Mongolia University of Technology. The building, which has the common functions, space and forms in 1960s, is present in the form of an industrial heritage. At present, however, we should reconsider its contemporary existence in reverse order.

Above all, I think that the inside space and outside space of the old industrial building are pretty consistent with each other, including the proportion of its spaces, their sizes, dimensions and division etc. It seems that the designers had taken the consistency to an extreme. What the building appeals to me is its another feature: it is conducive to communication between teachers and students. Students could be unconsciously influenced by the spaces and structures of the building and the techniques used in the construction of it as well. The dean of the department, Pengju, an intelligent man, has made the best use of the items in the building. It indicates that both teachers and students tend to take full advantage of the good environment in their teaching and learning of architecture. For example, what we can see in the building, such as the brick structure, wood structure, steel structure and the various nodes of the structures ect. is extremely helpful to our students' study of the course called Construction. The concepts are much easier to understand than in the traditional classroom where the teacher explains them out of thin air. It can be said that this is the best place for students to learn architecture because of the use of the most intuitive teaching approach. The building has great significance in education.

五、建筑馆的前世今生
V. The Past and Present Life of the Architectural Hall

1968
- 建设铸造车间。
- 提供教学实训。
- Built casting workshop.
- Provided the platform of teaching practice.

1978
- 投产，生产195型柴油机。
- Got into operation, produced the diesel engine.

1995
- 车间全面废弃。
- The workshop became deserted completely.

2008
- 建筑学部分师生向学校提出申请，保留改造厂房。
- 校党委决定改造为建筑馆，同时成立建筑学院。
- 改造工程开工。
- Teachers and students major in architecture filed a petition to the school to reserve and reconstruct the workshop.
- The CPC Committee of the university made the decision to reconstruct it to the Architectural Hall, and at the same time the College of Architecture was founded.
- The project of reconstructing came into operation.

2009
- 05月01日改造工程竣工、验收。
- 05月06日建筑馆迎接"全国建筑学专业评估委员会"专家。
- The reconstruction project was completed, checked and accepted on May 1.
- The Architectural Hall received the National Building Professional Assessment Committee on May 6.

2010
- 申请国家抗震经费，成立"内蒙古工业大学建筑馆抗震加固"课题组。对建筑馆实施研究性加固。
- 申请建筑馆扩建工程并获批。
- 06月获评为"内蒙古建筑设计一等奖。"
- 10月获评为"中国国际室内设计双年展金奖。"
- 申请改造建筑馆北侧的材料车间东侧部分为消防实训中心，纳入建筑馆的整体中。
- Filed a petition for National Earth Quake Funds, and established the research group of "Architecture Building Seismic Reinforcement in Inner Mongolia University of Technology" to investigative harden the Architectural Hall.
- Filed a petition to extend the Architectural Hall and got the permission.
- Won the first prize of Inner Mongolia Building Design in June.
- Was honored the "Gold Award for China International Biennale Interior Design" in October.
- Applied for rebuilding of the east part of the material workshop in the north of the Architectural Hall into Fire Fighting Training Center and brought the program into the whole of the Architectural Hall.

2011

- 06月由中国当代建筑创作论坛、《新建筑》杂志社、《城市·环境·设计》杂志社联合举办《新建筑·新求索》论坛，对建筑进行了现场品谈。
- 09月由建筑工业出版社出版《适应·更新·生长——一次人文与生态视野下的旧产业建筑改造实践》一书。
- 08月获评为"世界华人建筑师协会设计金奖。"
- 11月获评为"全国勘察设计行业一等奖。"
- The New Architecture and the New Quest Forum jointly organized by Chinese Contemporary Architectural Creation Forum, the magazine *New Architecture* and the magazine *City, Environment and Design* evaluated the building on spot in June.
- The book *Adjustment · improvement · Enlargement—An Old Factory Renovation in Views of Humanity and Ecology* was published by China Architecture & Building Press in September.
- Got the Gold prize for the World Chinese People Architects Association in August.
- Won the first award for the National Investigation and Designing in November.

2012

- 完成扩建工程设计方案，经学校审查通过，正式开工。
- 改造建筑馆西侧模型实验室为工业大学创新实验室。
- Had completed the plan of the extension project, examined and approved by the school and started.
- Transformed the prototyping laboratory in the west of the Architectural Hall into the Innovation Laboratory of Inner Mongolia University of Technology.

2013

- 02月建筑馆扩建工程竣工。同时一期工程中的教室部分改为教授和导师工作室。
- 05月迎接"全国建筑学评估委员会"复评专家。
- Completed the extension project of the Architectural Hall, meanwhile some classrooms in the first period were used as the studios for professors and supervisors in February.
- Received reassessment experts of the national building professional assessment committee in May.

2014

- 购置大型木制古建筑模型，安置在馆内外不同的空间内，增添了场所气氛。
- Purchased giant wooden ancient architectural model, placed at different spaces in and out of the building and added the workplace atmosphere.

2017

- 05月获得第十九届亚洲建筑协会建筑奖中的保护与改造类金奖。
- 05月迎接"堪培拉协议"国家观察专家接受专业教学评估。
- Win the gold award of Conservation and Renovation in the 19th ARCASIA awards for Architecture in May.
- Welcome the national observers of the "Canberra Agreement" and receive the specialty teaching evaluation in May.

2018

- 01月旧馆图书资料室并入学校图书馆，成为建筑学科馆，重新整修，并与沙龙打通，营造舒适的读书环境。
- 06月新馆一层东侧建设为"院士工作站"工作场所。
- The reference materials room of the college is integrated into the university library. The room becomes the Architecture Discipline Building, has been refurbished and integrated with the salon, creating comfortable reading environment in January.
- Take the east part of the first floor of the new building as an Academician Work Station in June.

图书在版编目（CIP）数据

适应·更新·生长 内蒙古工业大学建筑馆改扩建设计 /
张鹏举著 .— 北京：中国建筑工业出版社，2018.6
ISBN 978-7-112-22323-7

Ⅰ.①适… Ⅱ.①张… Ⅲ.①高等学校－教育建筑－改建－建筑设计－呼和浩特－汉、英②高等学校－教育建筑－扩建－建筑设计－呼和浩特－汉、英 Ⅳ.① TU244.3

中国版本图书馆 CIP 数据核字 (2018) 第 123690 号

本书立足于人文与生态视野，以内蒙古工业大学旧厂房改造实践为着手点，解析其设计理念与改造策略，尤其着眼于记录改造过程的点点滴滴。

本书适于建筑学师生、建筑设计从业者及爱好者参考阅读。

责任编辑：杨　晓　唐　旭　李东禧
责任校对：王　瑞
版式设计：扎拉根白尔
封面设计：李登钰
封面摄影：方振宁

适应·更新·生长
内蒙古工业大学建筑馆改扩建设计
张鹏举

*

中国建筑工业出版社出版、发行（北京海淀三里河路 9 号）
各地新华书店、建筑书店经销
北京雅昌艺术印刷有限公司印刷

*

开本：880×1230 毫米　1/16　印张：11 1/4　字数：408 千字
2018 年 7 月第一版　2018 年 7 月第一次印刷
定价：168.00 元
ISBN 978-7-112-22323-7
(32195)

版权所有　翻印必究
如有印装质量问题，可寄本社退换
（邮政编码 100037）